The IMA Volumes
in Mathematics
and Its Applications

Volume 1

Series Editors
George R. Sell   Hans Weinberger

Institute for Mathematics and Its Applications
IMA

The **Institute for Mathematics and Its Applications** was established by a grant from the National Science Foundation to the University of Minnesota in 1982. The IMA seeks to encourage the development and study of fresh mathematical concepts and questions of concern to the other sciences by bringing together mathematicians and scientists from diverse fields in an atmosphere that will stimulate discussion and collaboration.

The IMA Volumes are intended to involve the broader scientific community in this process.

Hans Weinberger, Director
George R. Sell, Associate Director

IMA Programs

1982–1983 Statistical and Continuum Approaches to Phase Transition

1983–1984 Mathematical Models for the Economics of Decentralized Resource Allocation

1984–1985 Continuum Physics and Partial Differential Equations

1985–1986 Stochastic Differential Equations and Their Applications

1986–1987 Scientific Computation

Summer 1987 Robotics

1987–1988 Applied Combinatorics

Springer Lecture Notes from the IMA

The Mathematics and Physics of Disordered Media
Editors: Barry Hughes and Barry Ninham
(Lecture Notes in Mathematics, Volume 1035, 1983)

Orienting Polymers
Editor: J. L. Ericksen
(Lecture Notes in Mathematics, Volume 1063, 1984)

New Perspectives in Thermodynamics
Editor: James Serrin
(in press)

Models of Economic Dynamics
Editor: Hugo Sonnenschein
(Lecture Notes in Economics, in press)

Homogenization and Effective Moduli of Materials and Media

Edited by
J. L. Ericksen, David Kinderlehrer,
Robert Kohn, and J.-L. Lions

With 41 Illustrations

Springer-Verlag
New York Berlin Heidelberg Tokyo

J. L. Ericksen
School of Mathematics and Department of Aerospace Engineering and Mechanics,
University of Minnesota, Minneapolis, MN 55455, U.S.A.
David Kinderlehrer
School of Mathematics, University of Minnesota, Minneapolis, MN 55455, U.S.A.
Robert Kohn
Courant Institute, New York University, New York, NY 10010, U.S.A.
J.-L. Lions
Centre National d'Etudes Spatiales, College de France, Paris 5, France
Institute for Mathematics and Its Applications
University of Minnesota, 514 Vincent Hall, 206 Church Street S.E., Minneapolis, MN
55455, U.S.A.

AMS Classification: 73BXX

Library of Congress Cataloging in Publication Data
Homogenization and effective moduli of materials and
  media.
  (IMA volumes in mathematics and its applications)
  Bibliography: p.
  1. Continuum mechanics.   2. Differential equations,
Partial. I. Ericksen, J.L. (Jerald L.), 1924–
II. Series
QA808.2.H66   1986   531   86–3927

Printed and bound by R.R. Donnelley & Sons, Harrisonburg, Virginia.
Printed in the United States of America.

9 8 7 6 5 4 3 2 1

ISBN 0-387-96306-5 Springer-Verlag New York Berlin Heidelberg Tokyo
ISBN 3-540-96306-5 Springer-Verlag Berlin Heidelberg New York Tokyo

The IMA Volumes in Mathematics and Its Applications

Current Volume:

Volume 1: Homogenization and Effective Moduli of Materials and Media
  Editors: Jerry Ericksen, David Kinderleherer, Robert Kohn, and J.-L. Lions

Forthcoming Volumes:

A
B
C
D
E

TABLE OF CONTENTS

FOREWORD

This IMA Volume in Mathematics and its Applications

Homogenization and Effective Moduli of Materials and Media

represents the proceedings of a workshop which was an integral part of the
1984-85 IMA program on CONTINUUM PHYSICS AND PARTIAL DIFFERENTIAL EQUATIONS.
We are grateful to the Scientific Committee:

J.L. Ericksen

D. Kinderlehrer

H. Brezis

C. Dafermos

for their dedication and hard work in developing an imaginative, stimulating, and
productive year-long program.

George R. Sell
Hans Weinberger

PREFACE

The papers in this volume were presented at a workshop on homogenization of differential equations and the determination of effective moduli of materials and media, primarily in the context of continuum theory. These areas are closely linked to a variety of phenomena, such as the elastic and dielectric responses of composites, and the effective properties of shales and soils. For instance, the ability to predict the effective stiffness response of a composite across a broad range of frequencies allows its performance under given circumstances to be assessed by means of nondestructive testing. A fundamental mathematical tool is homogenization, the study of partial differential equations with rapidly varying coefficients or boundary conditions. The recent alliance of homogenization with optimal design has stimulated the development of both fields. The presentations at the workshop emphasized recent advances and open questions.

The problem of effective moduli is an incompletely posed problem in the sense that the field equations governing the macroscopic body are not known at the outset but must be derived or approximated on the basis of the properties of its constitutents. The quest for a rational theory, its comparison with experiment, and the efficient reliable computation of solutions were primary concerns of this workshop. Such questions are among the principal themes of the 1984-1985 I.M.A. program, Continuum Physics and Partial Differential Equations.

The workshop brought together researchers in a number of areas of physics, engineering, and mathematics. The conference committee greatly appreciates the concerted efforts of the speakers and discussants to make their presentations intelligible to a mixed audience.

The conference committee would like to take this opportunity to thank the staff of the I.M.A., Professors Weinberger and Sell, Mrs. Susan Anderson, Mrs. Pat Kurth, and Mr. Robert Copeland, for their assistance in arranging the workshop. Special thanks are due to Mrs. Debbie Bradley and Mrs. Kaye Smith for their preparation of the manuscripts. We gratefully acknowledge the support of the National Science Foundation and the Office of Naval Research.

J.L. Ericksen

D. Kinderlehrer

R. Kohn

J.-L. Lions

conference committee

GENERALIZED PLATE MODELS AND OPTIMAL DESIGN

Martin P. Bendsøe
Mathematical Institute
The Technical University of Denmark
DK-2800 Lyngby, Denmark

Abstract

We consider the optimal design of linearly elastic, solid plates, that is, we seek the stiffest plate that can be made of a given amount of material. For large values of the ratio between the maximum allowable thickness and the minimum allowable thickness the stiffest plate cannot be obtained within the class of plates with slowly varying thickness; basically, this is caused by the cubic dependence of the rigidity tensor on the thickness function.

An extended class of plate models that allow for fields of stiffeners is described and it is shown how effective rigidity tensors can be obtained by homogenization or by a smear-out method based on continuity considerations. A computational optimization has been carried out and the numerical results indicate that use of this generalized plate model regularizes the optimization problem.

I. Introduction

Optimal design of linearly elastic plates presents an interesting and realistic case of the now well-established relationship between design and the study of materials with microstructure.

The problem of finding a thickness function for a solid, isotropic Kirchhoff plate that maximizes (for example) stiffness does not, in general, have a solution and plates with microstructure appear as natural relaxations of the original admissible type of plates. Numerical studies specifically lead one to consider plates stiffened with fields of infinitely many, infinitely thin integral ribs. In order to perform optimization within this set of generalized plate models effective moduli for these plates have to be computed.

The non-existence for the plate optimization problem for solid plates and the subsequent introduction of rib stiffened plates in the optimization problem was first studied by Cheng and Olhoff ([5], [6]) and Olhoff et. al ([17]) and the work presented here is a direct extension of their investigations.

II. The Plate Equation.

We consider a thin, solid elastic plate made of a linearly elastic material. The thickness h is variable and identifies the distance between the upper and lower plate surface, which are assumed to be disposed symmetrically with respect to the plate midplane. The plate midplane occupies a domain $\Omega \subseteq R^2$, so that the plate as a three-dimensional body occupies the domain $\Omega \times [\frac{-h}{2}, \frac{h}{2}]$ of R^3 . We shall use Kirchhoff plate theory, so our model is linear and deflections are assumed to be small. In a normal rectangular coordinate system in Ω the plate equation thus takes the form

$$\frac{\partial^2}{\partial x_\alpha \partial x_\beta} D_{\alpha\beta\kappa\gamma} \frac{\partial^2}{\partial x_\kappa \partial x_\gamma} w = p \quad \text{in } \Omega,$$

where w is the deflection of the plate correponding to the transverse load p , $\frac{\partial}{\partial x_\alpha}$ is the usual α -th partial derivative, $\alpha \in \{1,2\}$; Einstein conventions apply where indices appear. $e_{\alpha\beta} = \frac{\partial^2 u}{\partial x_\alpha \partial x_\beta}$ is the curvature tensor of the deformed plate. The tensor $D_{\alpha\beta\kappa\gamma}$ is the bending rigidity tensor for the plate and it depends on the geometry and material properties of the plate. $M_{\alpha\beta} = D_{\alpha\beta\kappa\gamma} e_{\kappa\gamma}$ is is the moment tensor of the plate. For a slowly varying thickness h and isotropic material the non-zero elements of $D_{\alpha\beta\kappa\gamma}$ are given by

$$\begin{aligned}
D_{\alpha\beta\kappa\gamma} &= h^3 \tilde{D}_{\alpha\beta\kappa\gamma} \\
\tilde{D}_{1111} = \tilde{D}_{2222} &= E/12(1 - \upsilon^2) = k \\
\tilde{D}_{1212} = \tilde{D}_{2121} = \tilde{D}_{1221} = \tilde{D}_{2112} &= (1 - \upsilon)k/2 \\
\tilde{D}_{2211} = \tilde{D}_{1122} &= \upsilon k
\end{aligned} \tag{2}$$

in which E and υ are Young's modulus and Poisson's ratio, respectively, for an isotropic, linearly elastic material.

The boundary conditions have not been specified and these will not play a significant role in this report. We will, though, always think of the homogeneous boundary conditions associated with a free, clamped or simply support edge. A completely free edge is not included so as to avoid rigid-body motions.

The weak, variational form of equation(1) is

$$w \in V \subseteq H^2(\Omega)$$
$$a_D(w,v) = b(v) \quad \text{for all} \quad v \in V \tag{3}$$

with

$$a_D(w,v) = \int_\Omega D_{\alpha\beta\kappa\gamma} \frac{\partial^2 w}{\partial x_\alpha \partial x_\beta} \frac{\partial^2 v}{\partial x_\kappa \partial x_\gamma} d\Omega \tag{4}$$
$$b(v) = \int_\Omega pv d\Omega$$

where V is a subspace of $H^2(\Omega)$ chosen in accordance with the imposed boundary conditions. Equation (3) is the virtual work equation, V is the set of kinematically admissable displacements, and v is a virtual displacement; $a_D(\, , \,)$ is the energy bilinear form and $b(\,)$ the load linear form. This variational equation makes it clear that we can consider the deflection w as an element of $H^2(\Omega)$, the load p as an element of $L^2(\Omega)$ and the elements $D_{\alpha\beta\kappa\gamma}$ of the rigidity tensor as elements of $L^\infty(\Omega)$.

For rigidities of the form (2) with a thickness function $h \in L^\infty(\Omega)$ satisfying $h \geq h_{min} > 0$, the symmetric form $a_D(\, , \,)$ is strongly elliptic, so there exists a $c > 0$, such that for $w \in V$:

$$a_D(w,w) \geq c\|w\|_{H^2}^2 \tag{5}$$

From this, existence of a unique solution to (3) follows, and this solution is bounded in $H^2(\Omega)$. We also have that the unique solution w of (3) depends continuously on the thickness h in L^∞ (see [9]), but this dependence is <u>not</u> linear and examples show that w does <u>not</u> depend continuously on h in L^∞-weak-*. Finally we note that the mapping $h \to w$ is Frechet differentiable ([9]).

III. The Design Problem.

The specific design problem we will consider is that of maximizing the stiffness of the plate using a given amount of material for its construction.

As a measure of stiffness we use the compliance (as is standard), which then has to be minimized in order to maximize stiffness. Inspired by the usual notation for optimum control problems, we write our problem as:

$$\text{minimize} \atop{D \epsilon U_{ad}} \quad \Pi(w) = \int_\Omega pw d\Omega$$

$$\text{subject to:} \quad \frac{\partial^2}{\partial x_\alpha \partial x_\beta} D_{\alpha\beta\kappa\gamma} \frac{\partial^2 w}{\partial x_\kappa \partial x_\gamma} = p, \tag{6}$$

$$\text{boundary conditions.}$$

The set $U_{ad} \subseteq (L^\infty(\Omega))^{16}$ is the set of admissible rigidity tensors (controls) for the problem, and we seek the minimum of the functional $\Pi(w) = \int_\Omega pw d\Omega$ on the state w, where the state w is related to the admissible controls via the plate equation and associated boundary conditions.

The natural choice for U_{ad} when considering plates ([15], [16]) is to choose the admissible rigidity tensors as arising from isotropic plates with slowly varying thickness, so that $D \epsilon U_{ad}$ has the form (cf. Eq. (2)):

$$D_{\alpha\beta\kappa\gamma} = h^3 \tilde{D}_{\alpha\beta\kappa\gamma}$$

with

$$0 < h_{min} \leqslant h \leqslant h_{max} \tag{7}$$

$$\int_\Omega h d\Omega = Vol \tag{8}$$

Here the constraints (7) are natural so as to avoid vanishing thicknesses (hinges, etc.) as well as very thin ribs of infinite height (see eq. [15], [16]). The constraint (8) is the constraint of the given volume, and in the following we use a plate of uniform thickness $h_u = Vol/Area(\Omega)$ as a reference plate of the given volume; we denote the compliance of this reference plate by Π_u. The optimization problem (6) with this type of rigidity tensors can then be written as

$$\text{minimize} \atop{h} \int_\Omega pw d\Omega$$

$$\text{subject to:} \quad a_D(w,v) = b(v) \quad \text{for all} \quad v \epsilon V \tag{9}$$

$$h_{min} \leqslant h \leqslant h_{max}$$

$$\int_\Omega h d\Omega = Vol$$

which is thus the problem of finding the thickness function h which produces the stiffest plate of a given volume.

The problem (9) cannot be solved analytically, but solutions can be found numerically by discretization using finite differences or finite elements. The optimization is performed iteratively, either based on a direct solution of the necessary conditions for optimality (optimality criteria method) or by employing well-known gradient type algorithms.

Computations show that the values of h_{min}, h_{max} and h_u chosen for the problem (9) play a very important role for the behaviour of the numerical solution that can be obtained.

For small spacings between h_{min}, h_u, h_{max} the problem seems to be well posed and different discretisations give rise to similar, seemingly well behaved designs, see Fig. 1.

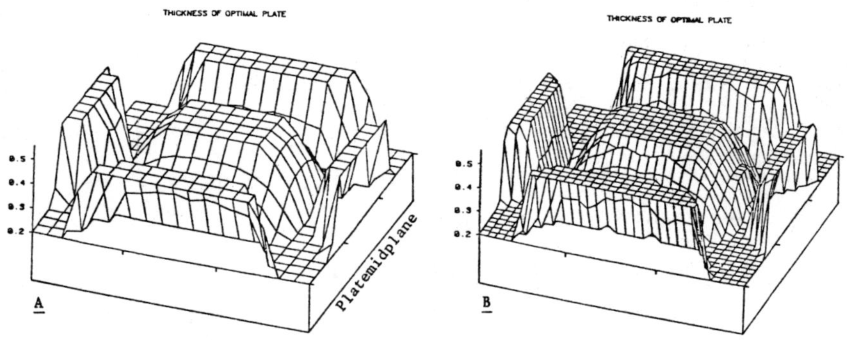

Fig. 1. Optimum (discretised) plate using the thickness as a design variable.
Values of constraints are h_{min} = 0.2 , h_{max} = 0.5 and h_u = 0.35.
A : for a 17 × 17 mesh, B : for a 30 × 30 mesh. Plate is clamped at the boundary.

However, for large values of h_{max} and small values of h_{min} and h_u the numerical solutions are not so well behaved. The solutions take on a stiffener-like structure, with stiffeners of height h_{max}, and an increase in the number of elements in the discretisation gives rise to a finer micro-structure and more stiffeners as well as to a substantial decrease in the value of the objective (the compliance); see Fig. 2.

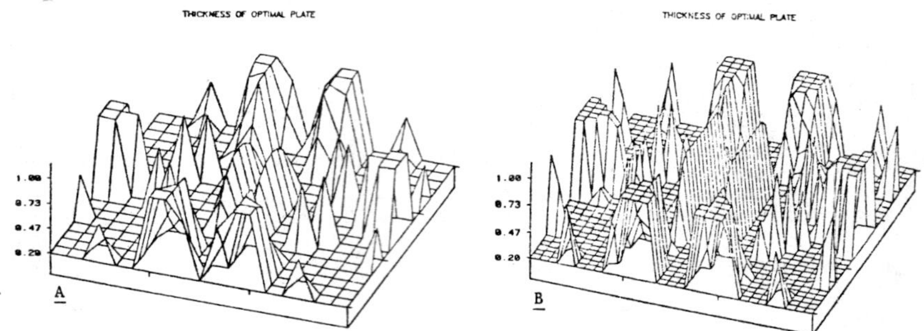

Fig. 2. Optimum (discretised) plate using the thickness as a designvariable. Values of constraints are h_{min} = 0.2 , h_{max} = 1.0 and h_u = 0.35 . A : for a 17 × 17 mesh, B : for a 30 × 30 mesh. Plate is clamped at the boundary.

This type of behavior was first studied in detail by Cheng and Olhoff ([5], Fig. 3) and their discovery has given rise to a substantial amount of research on plate optimization as well as optimum microstructure ([2]-[7], [11]-[14], [17], [18]; see also papers by Milton, Kohn, Vogelius, and Tartar in these proceedings).

The numerical experiments indicate that we cannot, in general, expect existence of solutions to the design problem (9) as formulated, using a slowly varying thickness. Also, the numerical work shows that it is more reasonable to consider the plate design problem (6) for an extended class of admissible rigidity tensors, containing (at least) plates consisting of a solid part of varying thickness as well as a number of thin, integral stiffeners in various directions and with varying densities (Fig. 3). The non-existence for large values of h_{max}/h_{min} can be shown analytically, as the admissible thicknesses for problem (9) in this case can fail to satisfy a Weierstrass necessary test for strong variations in a thin strip (see Refs. [12], [17]).

Basically, the non-existence for the problem (9) is caused by the fact that the set of admissible rigidity tensors depend on the cube of the thickness h: for tensors depending linearly on a parameter $t \in L^{\infty}(\Omega)$ that satisfy constraints of the form $\alpha \leqslant t \leqslant \beta$, $\int_{\Omega} t d\Omega = \Gamma$, existence for the minimum compliance problem would be assured ([3]). In order to identify more precisely the reason for non-existence, let

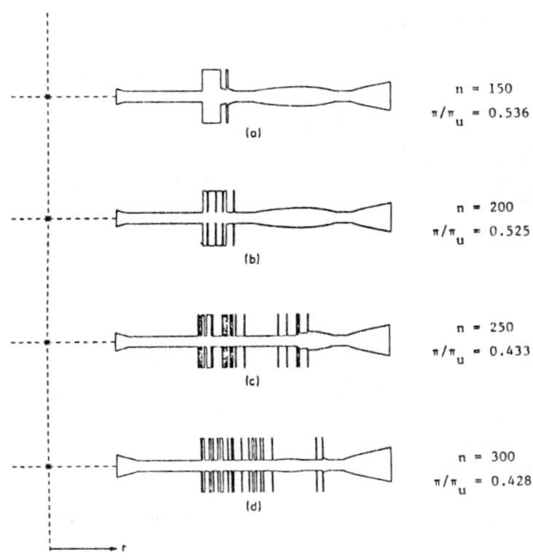

Fig. 3. The dependence of the optimum design of a doubly clamped
annular plate on the number n of elements used. The
plate is axisymmetric and subjected to a load of the form
p·cos4θ , in polar coordinates (r,θ) . From Ref. [5] .

$$W(U_{ad}) = \{w \; \varepsilon \; H^2(\Omega) \,|\, a_D(w,v) = b(v) \quad \text{all} \quad v, \text{ for some } D \; \varepsilon U_{ad}\} \qquad (10)$$

be the set of deflections corresponding to the admissible rigidities. For (9) the
constraint $0 < h_{min} < h$ implies that $W(U_{ad})$ is bounded in $H^2(\Omega)$ (cf. Eq.
(5)). Also, we see that the functional Π is weakly continuous on $H^2(\Omega)$, so the
reason for non-existence must thus be that $W(U_{ad})$ is <u>not</u> weakly closed in
$H^2(\Omega)$. This property means that any regularization of the plate design problem
(9) that should be performed in order to obtain existence must be carried out in a
way that is consistent with weak convergence of the corresponding deflections.
Thus one should seek a set U_{ad} of admissible designs which is closed in the
sense of G-convergence (also called H-convergence, i.e. convergence in the sense
of homogenization, see e.g. Ref. [19]), and which is a natural extension or
restriction of the original set of admissible tensors corresponding to plates of
slowly varying thickness. Such a G-closed set of designs not only assures
existence for the problem considered here but also for a whole class of plate

optimization problems with various other types of objectives ([4]).

One way ([3]) of obtaining a G-closed set of admissible designs for the optimization problem (9) is to consider only thickness functions h (for a plate with slowly varying thickness) that are uniformly bounded elements of $H^1(\Omega)$, i.e. with $M > 0$:

$$\int_\Omega (h^2 + (\text{grad } h)^2)d\Omega < M \tag{11}$$

By imposing this additional constraint we avoid the possibility of thin stiffeners with infinite slope and we see that the regularization is obtained by a variation of the well-known method of constraining (or penalizing) the value of the derivatives of the admissible controls (see also [14]). Using thicknesses in $H^1(\Omega)$ means that one restricts the design space. An extension of the design space, using relaxed controls, would involve finding the full G-closure of the set of rigidity tensors involved in (9) and it is not yet known how to do this for the general plate operator. For second order problems such as conductivity and torsion problems the G-closure can be found by employing a microstructure consisting of laminates with different scales. Such a microstructure could also be employed for a constant thickness plate which is made of two materials with either the same shear modulus or the same dilation modulus (see Refs. [11]-[13] and the contributions of Tartar and Kohn-Milton). In the following we shall describe an extended class of admissible plate types consisting of rib-stiffened plates. This class includes the limits of the minimizing sequences of stiffened plates obtained from the numerical computations of the optimal plate with a slowly varying thickness, cf. Figs. 2 and 3.

IV. Ribstiffened Plates. Effective Moduli.

The rib stiffened structure of the optimal designs obtained from a discretised version of problem (a), cf. Figs. 2 and 3, makes it natural to consider optimization within a generalized class of plate models consisting of plates stiffened with fields of infinitely many, infinitely thin stiffeners. This was first studied by Cheng, Cheng and Olhoff, and Olhoff, Lurie et. al ([6],[7],[17]) for axisymmetric plates, where the symmetry implies that there is just one field of stiffeners running circumferentially around the plate. This work has led to the general study of and use of laminates for various design problems ([11]-[13]) as well as to a study of rib-stiffened plastic plates ([18]). We shall here describe a similar model for general, two-dimensional plates.

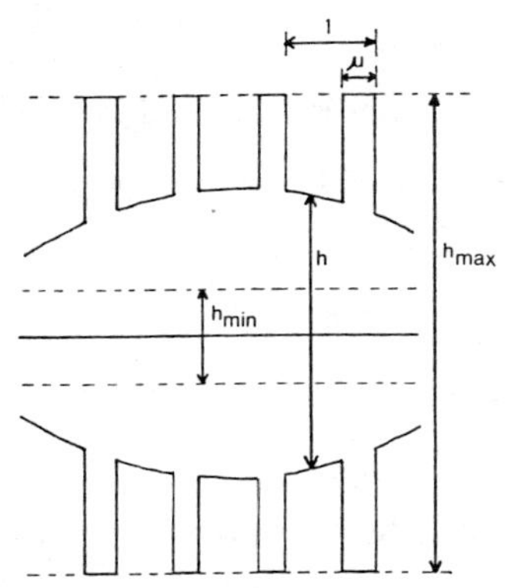

Fig. 4. Cross-section of rib-stiffened plate with one field of stiffeners of density μ .

The plate models we shall consider are two-dimensional versions of the models used in Refs. [6], [7] and [17]. The plates consist of a solid part of slowly varying thickness h that is constrained, $h_{min} \leqslant h \leqslant h_{max}$, as well as two mutually orthogonal fields of infinitely many, infinitely thin stiffeners of height h_{max}; see Fig. 4. The number of stiffeners in a field is characterized by a density μ so that in a unit cell a stiffener of a field of density μ has the width μ (see fig. 5).

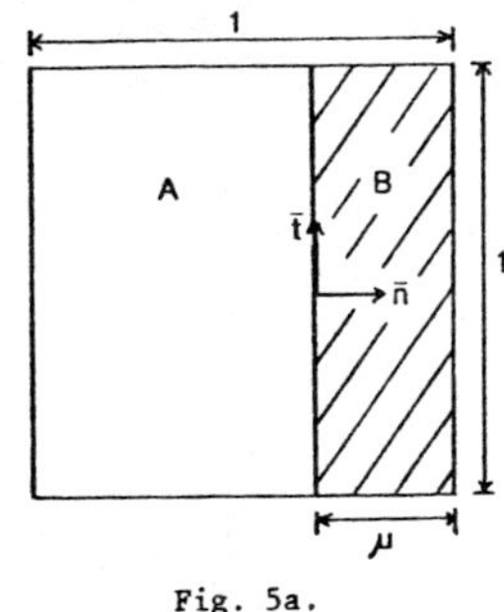

Fig. 5a.

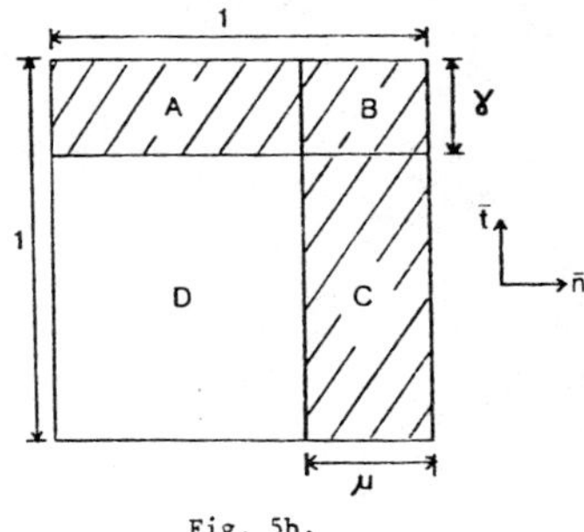

Fig. 5b.

Fig. 5. Unit cell for ribstiffened plate with one field of stiffeners of density μ (5a) and unit cell for ribstiffened plate with two fields of stiffeners with densities μ and γ (5b).

The unit cell for one field of stiffeners is shown in Fig. 5a; there the direction of the field is given by the vector $\bar{t}$. The basic unit cell for two orthogonal fields of stiffeners is shown in Fig. 5b; there the density of the field in the direction $\bar{t}$ is μ and the density of the field in the direction $\bar{n}$

is γ. We see that the for this new plate model the form of the plate is described by four functions in the plate domain Ω namely the thickness h, the two densities μ and γ and an angle θ giving $\overline{n} = (\cos\theta, \sin\theta)$ and $\overline{t} = (-\sin\theta, \cos\theta)$ in a fixed coordinate system. Optimization within this extended class requires optimization with respect to all these distributed design variables. Since in the original optimization problem we used the Kirchhoff plate equation and, since we wish to extend the design space for the problem (9) and not to introduce a new problem with altered state equation, we will treat these generalized plates as Kirchhoff plates and will take the micsrostructure into account by using effective rigidity tensors in the state equation (1).

It was noted earlier that the functional Π in the optimization problem (9) is weakly continuous in the deflection w and that any extension of the set of admissible rigidity tensors should be made so that the corresponding deflection stays within the weak closure of the set $W(U_{ad})$ of deflections corresponding to slowly varying thicknesses. This makes it natural to obtain the effective moduli by homogenization, i.e. we consider the ribstiffened plates as limits (for $\varepsilon \to 0$) of plates with rigidities $h(x, \frac{x}{\varepsilon})^3 \tilde{D}_{\alpha\beta\kappa\gamma}$, $x \in \Omega$, where $h(x,y)$ is periodic in y. The limit is taken so that the corresponding deflection w for the homogenized tensor is the weak limit of the deflections w_ε ([19]). For the plate equation (1) the effective moduli $q_{\alpha\beta\kappa\gamma}$ are given by ([8]):

$$q_{\alpha\beta\kappa\gamma} = \frac{1}{|Y|} a_Y(\chi^{\alpha\beta} - y^{\alpha\beta}, \chi^{\kappa\gamma} - y^{\kappa\gamma}), \qquad (12)$$

where

$$y^{\alpha\beta} = \begin{cases} \frac{1}{2} x_1^2 \ , & \alpha\beta = 11, \\[2mm] \frac{1}{2} x_1 x_2, & \alpha\beta = 12 \text{ or } 21, \\[2mm] \frac{1}{2} x_2^2 \ , & \alpha\beta = 22. \end{cases} \qquad (13)$$

a_Y is the plate bilinear form (cf. Eq. (4)) on the unit period cell Y (with area $|Y|$) and $\chi^{\alpha\beta}$ is a periodic solution of the following plate equation in the unit cell:

$$a_Y(\chi^{\alpha\beta} - y^{\alpha\beta}, \psi) = 0 \quad \text{for all periodic} \quad \psi \in H^2(Y). \tag{14}$$

For a plate with one field of stiffeners in the direction of the x_2-axis and with a density μ, the unit cell Y is an infinite strip (Fig. 6) and solving (14) reduces to solving a one-dimensional problem.

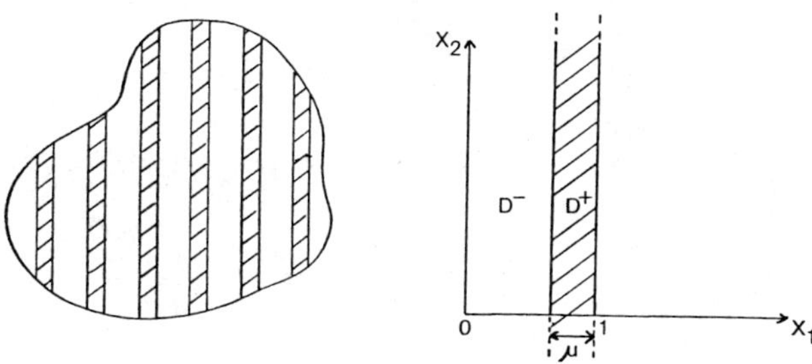

Fig. 6. Homogenization of a plate with one field of stiffeners of density μ.

The periodic thickness function is the periodic extension of

$$\overline{h}(x,y) = \begin{cases} h(x_1) & 0 < y_1 < \mu(x_1) \\[2mm] h_{max} & \mu(x_1) < y_1 < 1 \end{cases} \tag{15}$$

so that the unstiffened region in the unit cell has a rigidity tensor $D^- = h^3\tilde{D}$ and the stiffened region has a rigidity tensor $D^+ = h_{max}^3\tilde{D}$. The following calculations then give $q_{\alpha\beta\kappa\gamma}$ (we write D for the element D_{1111} of the rigidity tensor $\overline{h}^3\tilde{D}$):

$$\text{A)} \quad a_Y(\chi^{11}, \psi) = \int_Y \frac{\partial^2 D_{11\kappa\gamma}}{\partial x_\kappa \, \partial x_\gamma} \psi \, dY \quad \text{for all} \quad \psi,$$

so $D_{1111} \dfrac{\partial^2 \chi^{11}}{\partial x_1^2} = D_{1111} + c_1$. Periodicity gives $c_1 = \dfrac{-1}{M(\frac{1}{D})}$ and thus

$$q_{1111} = \frac{1}{M(\frac{1}{D})} \, ,$$

where $M(f)$ denotes the mean of f over the cell Y.

12

B) $\quad a_\gamma(\chi^{22}, \psi) = \int_Y \dfrac{\partial^2 D_{22\kappa\gamma}}{\partial x_\kappa \, \partial x_\gamma} \, \psi \, dY \quad$ for all $\quad \psi$,

so $\quad D_{1111} \dfrac{\partial^2 \chi^{22}}{\partial x_1^{\,2}} = D_{1122} + c_2;$ Periodicity gives $\quad c_2 = \dfrac{-\upsilon}{M\left(\frac{1}{D}\right)}$,

and thus $\quad q_{2222} = M(D) - \upsilon^2 M(D) + \upsilon^2 \dfrac{1}{M\left(\frac{1}{D}\right)}$.

C) $\quad a_\gamma(\chi^{12}, \psi) = 0 \quad$ for all $\quad \psi$, so $\quad D_{1111} \dfrac{\partial^2 \chi^{12}}{\partial x_1^{\,2}} = c_3.$

Periodicity gives $\quad c_3 = 0$, and thus $\quad q_{1212} = \dfrac{(1 - \upsilon)}{2} M(D).$

D) From A) we have χ^{11}, and from this

$$q_{1122} = \upsilon \, \dfrac{1}{M(D)} \, .$$

Calculating $M(D)$ and rearranging leads to the following expression for the effective moduli:

$$q_{1111} = \dfrac{D^+ D^-}{\mu D^- + (1 - \mu)D^+} \, ,$$

$$q_{2222} = \mu D^+ + (1 - \mu)D^- - \dfrac{\upsilon^2(1 - \mu)\mu}{\mu D^- + (1 - \mu)D^+} \, (D^+ - D^-)^2,$$

$$q_{1212} = \dfrac{1 - \upsilon}{2} \, (\mu D^+ + (1 - \mu)D^-),$$

$$q_{1122} = \upsilon \, \dfrac{D^+ D^-}{\mu D^- + (1 - \mu)D^+} \, ,$$

so the homogenized material is orthotropic, but anisotropic, with planes of symmetry determined by the direction of the stiffener. With the stiffener in any direction $\bar{t}$ (cf. Fig. 5), the formula (16) has the tensorial form:

$$q_{\alpha\beta\kappa\gamma} = (1 - \mu)D^-_{\alpha\beta\kappa\gamma} + \mu \, D^+_{\alpha\beta\kappa\gamma}$$

$$- \dfrac{\mu(1 - \mu)}{\bar{D}} \, (D^-_{\alpha\beta\epsilon\theta} - D^+_{\alpha\beta\epsilon\theta})(D^-_{\zeta\eta\kappa\gamma} - D^+_{\zeta\eta\kappa\gamma}) n_\zeta n_\theta n_\epsilon n_\theta; \quad (17)$$

$$\bar{D} = ((1 - \mu)D^+_{\alpha\beta\kappa\gamma} + \mu D^-_{\alpha\beta\kappa\gamma}) n_\alpha n_\beta n_\kappa n_\gamma.$$

The coefficients $q_{\alpha\beta\kappa\gamma}$ of Eq. (17) can also be calculated using an averaging principle and certain continuity conditions along the interfaces between the unstiffened zone A and the stiffened zone B (Fig. 5a). This was used in Refs. [6], [7] and [17] for the study of optimal design of axially symmetric plates with one field of circumferential stiffeners; the optimization results of these papers show that rib-stiffened plates are indeed superior and the results also indicate that the extended design space regularizes the problem.

For a plate with two fields of stiffeners, an analytical homogenization cannot be performed, as it is not possible to give an analytical solution of the plate equation for the unit cell of this structure. To optimize the thickness h, the densities μ, γ, and direction of the stiffeners using the homogenized state operator would thus require, at each discretisation point, a numerical computation of the homogenized coefficients. Moreover, this would have to be performed at each step of an iterative optimization algorithm. Alternatively one could compute a table of effective moduli and use this table in the optimization. Either procedure would of course be very expensive if a reasonable numerical accuracy is to be achieved. We shall therefore take a different approach, by deriving an analytical expression that approximates the effective moduli, using an averaging principle. For one field of stiffeners the same principle leads to the exact effective moduli.

For the plate with two fields of stiffeners we have a unit cell as shown in Fig. 5b. In each of the regions of the element there is a moment-curvature relationship

$$M^i = D^+_{\alpha\beta\kappa\gamma} e^i_{\kappa\gamma} \ , \quad i = A, B, D,$$

$$M^C = D^-_{\alpha\beta\kappa\gamma} e^C_{\kappa\gamma} \ . \tag{18}$$

Our ansatz is that, in the limit, the stiffened plate can be described by the averaged tensors

$$e = \gamma(1 - \mu)e^A + \mu(1 - \gamma)e^B + (1 - \mu)(1 - \gamma)e^C + \mu\gamma e^D, \tag{19}$$

$$M = \gamma(1 - \mu)M^A + \mu(1 - \gamma)M^B + (1 - \mu)(1-\gamma)M^C + \mu\gamma M^D, \tag{20}$$

and the average bending rigidity tensor $D_{\alpha\beta\kappa\gamma}$ of the rib-stiffened plate is then the tensor (if it exists) that gives the relation

$$M_{\alpha\beta} = D_{\alpha\beta\kappa\gamma} \, e_{\kappa\gamma}. \tag{21}$$

We have thus assumed that the stiffened plate can be described by the standard plate equation and that the curvature and moment tensors for this description are the averages in a unit cell of the curvature and moment tensors in each of the separate regions A,B,C and D. To calculate the effective tensor $D_{\alpha\beta\kappa\gamma}$ we can use that certain continuity conditions must hold for the moments and curvatures along the interfaces of the unit cell. In order to obtain the correct formula (17) for the case of one field of stiffeners, we assume continuity of the tangential components of the curvatures and of the normal component of the moment along the interfaces between the stiffened regions A, C and the unstiffened region D, while along the interfaces in the stiffeners we only assume continuity of the curvatures. These choices are based on physical considerations as well as on variational arguments. We remark that other combinations of continuity conditions together with the ansatz lead to physically unreasonable effective moduli (e.g. conditions like $\upsilon = 0$).

The continuity conditions at the interfaces between the parts of the basic cell give for the mixed, tangential components of the curvature that

$$e^A_{\alpha\beta} t_\alpha n_\beta = e^C_{\alpha\beta} t_\alpha n_\beta = e^B_{\alpha\beta} t_\alpha n_\beta = e^D_{\alpha\beta} t_\alpha n_\beta. \tag{22}$$

Combining (19) and (22) gives

$$e^I_{\alpha\beta} = e_{\alpha\beta} + I_1 n_\alpha n_\beta + I_2 t_\alpha t_\beta, \quad I = A,B,C,D, \tag{23}$$

with constants A_α, B_α, C_α, D_α, $\alpha = 1,2$. Continuity of the tangential components of the curvature give for the interfaces between A and C and B and D respectively:

$$e^A_{\alpha\beta} n_\alpha n_\beta = e^C_{\alpha\beta} n_\alpha n_\beta \; ; \; e^B_{\alpha\beta} n_\alpha n_\beta = e^D_{\alpha\beta} n_\alpha n_\beta , \tag{24}$$

which means (see (23)) that $A_1 = C_1$ and $B_1 = D_1$. Similarly we have that

$$e^A_{\alpha\beta} t_\alpha t_\beta = e^D_{\alpha\beta} t_\alpha t_\beta; \quad e^B_{\alpha\beta} t_\alpha t_\beta = e^C_{\alpha\beta} t_\alpha t_\beta, \tag{25}$$

so that $A_2 = D_2$ and $B_2 = C_2$. Thus (23) reduces to

$$e^A_{\alpha\beta} = e_{\alpha\beta} + A_1 n_\alpha n_\beta + A_2 t_\alpha t_\beta,$$

$$e^B_{\alpha\beta} = e_{\alpha\beta} + B_1 n_\alpha n_\beta + B_2 t_\alpha t_\beta,$$

$$e^C_{\alpha\beta} = e_{\alpha\beta} + A_1 n_\alpha n_\beta + B_2 t_\alpha t_\beta \tag{26}$$

$$e^D_{\alpha\beta} = e_{\alpha\beta} + B_1 n_\alpha n_\beta + A_2 t_\alpha t_\beta.$$

Inserting (26) into (19) and contracting over $n_\alpha n_\beta$ or $t_\alpha t_\beta$ this gives that

$$(1 - \mu)A_1 + \mu B_1 = 0 \quad \text{and} \quad \gamma A_2 + (1 - \gamma)B_2 = 0. \tag{27}$$

We now turn to the continuity in moments, which are the conditions:

$$M^B_{\alpha\beta} n_\alpha n_\beta = M^C_{\alpha\beta} n_\alpha n_\beta \; ; \; M^A_{\alpha\beta} t_\alpha t_\beta = M^C_{\alpha\beta} t_\alpha t_\beta. \tag{28}$$

From (18) and (26) these equations imply that

$$D^+_{\alpha\beta\kappa\gamma}(e_{\kappa\gamma} + B_1 n_\kappa n_\gamma + B_2 t_\kappa t_\gamma)n_\alpha n_\beta = \tag{29}$$
$$D^-_{\alpha\beta\kappa\gamma}(e_{\kappa\gamma} + A_1 n_\kappa n_\gamma + B_2 t_\kappa t_\gamma)n_\alpha n_\beta \, ,$$

and

$$D^+_{\alpha\beta\kappa\gamma}(e_{\kappa\gamma} + A_1 n_\kappa n_\gamma + A_2 t_\kappa t_\gamma)t_\alpha t_\beta =$$
$$D^-_{\alpha\beta\kappa\gamma}(e_{\kappa\gamma} + A_1 n_\kappa n_\gamma + B_2 t_\kappa t_\gamma)t_\alpha t_\beta. \tag{30}$$

Now multiply (29) and (30) with $\mu(1 - \gamma)$, rearrange and use (27) to obtain:

$$A_1[(1-\mu)(1-\gamma)D^+ + \mu(1-\gamma)D^-]_{\alpha\beta\kappa\gamma} n_\kappa n_\gamma n_\alpha n_\beta - \mu\gamma A_2[D^- - D^+]_{\alpha\beta\kappa\gamma} n_\alpha n_\beta t_\kappa t_\gamma \tag{31}$$
$$= \mu(1 - \gamma)[D^+ - D^-]_{\alpha\beta\kappa\gamma} e_{\kappa\gamma} n_\alpha n_\beta,$$

and

$$\mu(1-\gamma)A_1[D^- - D^+]_{\alpha\beta\kappa\gamma} t_\alpha t_\beta n_\kappa n_\gamma - A_2[\mu(1-\gamma)D^+ + \mu\gamma D^-]_{\alpha\beta\kappa\gamma} t_\alpha t_\beta t_\kappa t_\gamma \tag{32}$$
$$= \mu(1-\gamma)[D^+ - D^-]_{\alpha\beta\kappa\gamma} e_{\kappa\gamma} t_\alpha t_\beta.$$

Using now the expressions

$$E_1 = [(1 - \gamma)D^+ + \gamma D^-]_{\alpha\beta\kappa\gamma} \, t_\alpha t_\beta t_\kappa t_\gamma,$$

$$E_2 = [(1 - \mu)D^+ + \mu D^-]_{\alpha\beta\kappa\gamma} \, n_\alpha n_\beta n_\kappa n_\gamma,$$

$$E_3 = [D^- - D^+]_{\alpha\beta\kappa\gamma} \, n_\alpha n_\beta t_\kappa t_\gamma,$$

$$\overline{D} = -E_1 E_2 + \mu\gamma E_3^2,$$

formulas (31) and (32) can be written as:

$$A_1(1 - \gamma)E_2 - A_2 \, \mu\gamma \, E_3 = \mu(1 - \gamma)[D^+ - D^-]_{\alpha\beta\kappa\gamma} e_{\kappa\gamma} n_\alpha n_\beta, \tag{33}$$

and

$$A_1\mu(1 - \gamma)E_3 - A_2 \, \mu E_1 = \mu(1 - \gamma)[D^+ - D^-]_{\alpha\beta\kappa\gamma} e_{\kappa\gamma} t_\alpha t_\beta, \tag{34}$$

and thus we have two linear equations in the two unknowns A_1, A_2 and the determinant of these equations is $\mu(1 - \gamma)\overline{D}$. Solving, we have

$$A_1 = \frac{1}{\overline{D}} \, [-\mu \, E_1 [D^+ - D^-]_{\alpha\beta\kappa\gamma} e_{\kappa\gamma} n_\alpha n_\beta + \mu\gamma \, E_3 [D^+ - D^-]_{\alpha\beta\kappa\gamma} e_{\kappa\gamma} t_\alpha t_\beta], \tag{35}$$

and

$$A_2 = \frac{1}{\overline{D}} \, [(1-\gamma)E_2 [D^+ - D^-]_{\alpha\beta\kappa\gamma} e_{\kappa\gamma} t_\alpha t_\beta - \mu(1-\gamma)E_3 [D^+ - D^-]_{\alpha\beta\kappa\gamma} e_{\kappa\gamma} n_\alpha n_\beta]. \tag{36}$$

We now have the curvative tensors e^I, $I = A,B,C,D$ given entirely by the averaged curvature tensor e (combine (26), (27) and (35), (36)). The equation for the average momentum tensor M given by (20) thus reduces to

$$\begin{aligned}
M_{\alpha\beta} &= (1 - \gamma)D^+_{\alpha\beta\kappa\gamma} e_{\kappa\gamma} + \lambda D^-_{\alpha\beta\kappa\gamma} e_{\kappa\gamma} \\
&\quad - (1 - \mu)(1 - \gamma)A_1 [D^+ - D^-]_{\alpha\beta\kappa\gamma} n_\kappa n_\gamma \\
&\quad + \gamma(1 - \mu)A_2 [D^+ - D^-]_{\alpha\beta\kappa\gamma} t_\kappa t_\gamma,
\end{aligned} \tag{37}$$

by use of the moment-curvative relations given in (18). Inserting the expressions (35) and (36) for A_1 and A_2 into this equation then leads to the sought desired relation

$$M_{\alpha\beta} = D_{\alpha\beta\kappa\gamma} e_{\kappa\gamma},$$

with $D_{\alpha\beta\kappa\gamma}$ given by:

$$D_{\alpha\beta\kappa\gamma} = (1 - \lambda)D^+_{\alpha\beta\kappa\gamma} + \lambda D^-_{\alpha\beta\kappa\gamma} \qquad (38)$$

$$+ \mu(1-\mu)(1-\gamma)\frac{E_1}{\overline{D}}(D^+ - D^-)_{\alpha\beta\epsilon\theta}n_\epsilon n_\theta(D^+ - D^-)_{\zeta\eta\kappa\gamma}n_\zeta n_\eta$$

$$+ \gamma(1-\gamma)(1-\mu)\frac{E_2}{\overline{D}}(D^+ - D^-)_{\alpha\beta\epsilon\theta}\,t_\epsilon t_\theta(D^+ - D^-)_{\zeta\eta\kappa\gamma}\,t_\zeta t_\eta$$

$$- \mu(1-\mu)(1-\gamma)\gamma\frac{E_3}{\overline{D}}(D^+ - D^-)_{\alpha\beta\epsilon\theta}\,n_\epsilon n_\theta(D^+ - D^-)_{\zeta\eta\kappa\gamma}\,t_\zeta t_\eta$$

$$- \mu(1-\mu)(1-\gamma)\gamma\frac{E_3}{\overline{D}}(D^+ - D^-)_{\alpha\beta\epsilon\theta}t_\epsilon t_\theta(D^+ - D^-)_{\zeta\eta\kappa\gamma}n_\zeta n_\eta,$$

with

$$\lambda = (1 - \mu)(1 - \gamma)\,,$$

$$E_1 = [(1 - \gamma)D^+ + \gamma D^-]_{\alpha\beta\kappa\gamma}\,t_\alpha t_\beta t_\kappa t_\gamma,$$

$$E_2 = [(1 - \mu)D^+ + \mu D^-]_{\alpha\beta\kappa\gamma}\,n_\alpha n_\beta n_\kappa n_\gamma,$$

$$E_3 = [D^- - D^+]_{\alpha\beta\kappa\gamma}\,n_\alpha n_\beta t_\kappa t_\gamma,$$

$$\overline{D} = -E_1 E_2 + \mu\gamma E_3^2\,.$$

It should be pointed out that the derivation of (38) is only reasonable for
values of $\mu + \gamma$ less than 1 or for a Poisson ratio υ equal to zero. For
large values of μ, γ, υ and D^+/D^- the determinant $\overline{D}$ for the linear equations
(33), (34) can become zero and the method breaks down. For the optimization
results to be reported this does not, however, play any role, as we are considering small volume fractions, i.e. small values of h_u/h_{max}; this automatically
forces $\mu + \gamma$ to be small. Also a few computations of homogenized moduli for the
two-way ribstiffened plates have been carried out for small values of μ and γ
and the results compare very well with the results obtained using (38).

The smeared-out tensor given by (38) has the desired property that in the
limits $(\mu,\gamma) = (0,0)$ and $\mu = 1$ or $\gamma = 1$ it reduces to the correct tensor for
an isotropic, solid plate of thickness h and h_{max}, respectively. In the case
of one density being zero and the other density being non-zero the tensor reduces
to the correct one (17) for a plate with one field of stiffeners. Also for small
values of γ and $\gamma \ll \mu$ the limit of (38) is the same as that obtained by
treating the plate as "a laminate of laminates", which means that the plate is

regarded as a laminate of a solid material of density μ with some other material (of density $(1 - \mu)$) which is itself a laminate with stiffeners of density γ. (The idea of laminates of laminates is discussed in detail by Milton [these proceedings].)

We note that a material described by the rigidity tensor (38) is anisotropic but orthotropic and the planes of symmetry are given by the directions of the stiffeners.

The application of homogenization to a Kirchhoff plate model in the case of a rib-stiffened plate is from a physical point of view somewhat contradictory, as the Kirchhoff plate model assumes that the thickness is slowly varying! When considering plate optimization with thickness as the design variable it is, however, necessary to consider the rib-stiffened case in order to obtain the optimal performance. Also practical designs can be obtained by a lumping process (cf. [7]) and for this purpose homogenization of the Kirchhoff equations is the correct procedure. For a general study of optimization of plates with rapidly varying thickness, new plate models can be derived from 3-dimensional elasticity, as shown in Ref. [10] (for some cases, the three-dimensional model gives the same equations as homogenization of the Kirchhoff equations). As these general models for plates with rapidly varying thickness also cannot be given an analytical form when variation in the thickness occurs in more than one direction, optimization of plates modelled in this way requires considerable computational effort. An alternative to this improvement of the modelling is to treat the stiffeners as beam elements. This is perhaps more realistic, as a stiffener in this case does not add to the plate any torsional rigidity in its own direction; and for a plate stiffened with one or two fields of infinitely many, infinitely thin beams, an analytical expression of the effective rigidity can be obtained via homogenization ([1]). Introducing beam elements does change the basic model, but it is interesting to compare the resulting optimal designs originating from the two approahces. If the beams are considered to have no torsional stiffness, then homogenization based on the variational principle leads to an effective rigidity tensor $\hat{D}_{\alpha\beta\kappa\gamma}$ given by superposition:

$$\hat{D}_{\alpha\beta\kappa\gamma} = h^3 \tilde{D}_{\alpha\beta\kappa\gamma} + \frac{E}{12}(h^3_{max} - h^3)(\gamma n_\alpha n_\beta n_\kappa n_\gamma + \mu t_\alpha t_\beta t_\kappa t_\gamma) \tag{39}$$

where the beam density is γ in the direction $\overline{n}$ and μ in the direction $\overline{t}$.

V. The Optimal Designs

We now consider the problem of optimizing plates equipped with two mutually orthogonal fields of infinitely many, infinitely thin integral stiffeners. The design variables are the variable thickness h of the solid part of the plate, the densities μ, γ , and the directions (given by an angle θ) of the two fields of stiffeners:

$$\begin{array}{ll} \underset{h,\mu,\gamma,\theta}{\text{minimize}} & \int_\Omega \rho w d\Omega \\[2ex] \text{subject to:} & a_D(w,v) = b(v) , \quad \text{all} \quad v \in V \\[1ex] & h_{min} \leqslant h \leqslant h_{max} \\[1ex] & 0 \leqslant \mu \leqslant 1, \; 0 \leqslant \gamma \leqslant 1. \\[1ex] & \int_\Omega ((1-\mu)(1-\gamma)h + (\mu + \gamma - \mu\gamma)h_{max})d\Omega = \text{Vol} \end{array} \tag{40}$$

where the rigidities D are given by (38). Necessary condition for optimality can be obtained by taking first variations of the augmented Lagrangian

$$\begin{aligned} L = & \int_\Omega \rho w d\Omega + [a_D(w,\overline{w}) - b(\overline{w})] \\ & + \int_\Omega [n_1(h_{min} - h + \sigma_1^2) + (h - h_{max} + \sigma_1^2)n_2 \\ & + \alpha_1(-\mu + \sigma_3^2) + \alpha_2(\mu - 1 + \sigma_4^2) \\ & + \beta_1(-\gamma + \sigma_5^2) + \beta_2(\gamma - 1 + \sigma_6^2) \\ & + \Lambda((1-\mu)(1-\gamma)h + (\mu + \gamma - \mu\gamma)h_{max} - \text{Vol})]d\Omega, \end{aligned} \tag{41}$$

where $\overline{w}$ is a Lagrange multiplier function associated with the equality constraint $(D,w) \rightarrow a_D(w,\cdot) - b(\cdot)$, Λ is a Lagrangian multiplier associated with the volume constraint, and $\eta_i, \alpha_i, \beta_i$, $i = 1, 2$, are Lagrange multipliers associated with the constraints on the design variables. The σ_i, $i = 1, \ldots, 6$ are slack variables that convert the inequality constraints into equality constraints.

Variations of L with respect to w give that $\overline{w} = -w$, and variations of L with respect to the design variables give the necessary conditions:

$$\frac{\partial D_{\alpha\beta\kappa\gamma}}{\partial h} e_{\alpha\beta} e_{\kappa\gamma} = \Lambda(1 - \mu)(1 - \gamma) + \eta_1 - \eta_2, \tag{42a}$$

$$\frac{\partial D_{\alpha\beta\kappa\gamma}}{\partial \mu} e_{\alpha\beta} e_{\kappa\gamma} = \Lambda(1 - \gamma)(h_{max} - h) + \alpha_1 - \alpha_2, \tag{42b}$$

$$\frac{\partial D_{\alpha\beta\kappa\gamma}}{\partial \gamma} e_{\alpha\beta} e_{\kappa\gamma} = \Lambda(1 - \mu)(h_{max} - h) + \beta_1 - \beta_2, \tag{42c}$$

$$\frac{\partial D_{\alpha\beta\kappa\gamma}}{\partial \theta} e_{\alpha\beta} e_{\kappa\gamma} = 0, \tag{42d}$$

where $e_{\alpha\beta} = \dfrac{\partial^2 w}{\partial x_\alpha \partial x_\beta}$ are the curvatures. First and second order necessary conditions for a minimum taken with respect to the slack variables lead to the switching conditions:

$$\begin{aligned}
\alpha_1 &= 0, \ \alpha_2 \geqslant 0 \quad \text{if} \ \ \mu = 0, \\
\alpha_1 &= 0, \ \alpha_2 = 0 \quad \text{if} \ \ 0 < \mu < 1, \\
\alpha_1 &\geqslant 0, \ \alpha_2 = 0 \quad \text{if} \ \ \mu = 1,
\end{aligned} \tag{43}$$

Likewise for $(\beta_1, \beta_2, \gamma)$, and for η_i we have:

$$\begin{aligned}
\eta_1 &= 0, \ \eta_2 \geqslant 0 \quad \text{if} \ \ h = h_{min}, \\
\eta_1 &= 0, \ \eta_2 = 0 \quad \text{if} \ \ h_{min} < h < h_{max}, \\
\eta_1 &\geqslant 0, \ \eta_2 = 0 \quad \text{if} \ \ h = h_{max},
\end{aligned} \tag{44}$$

In the problem formulation above we have assumed a priori that the two fields of stiffeners in the optimal plate must be orthogonal. This intuitive hypothesis has been shown correct for the design of <u>plastic</u> plates ([18]). In the elastic case (with effective rigidities given by (38)) one can show that having the stiffeners directed along the (orthogonal) principal directions of the moment tensor is a stationary situation, i.e. it satisfies a first order optimality condition

We used the necessary conditions (42)-(44) to calculate optimal designs for given values of load p and constraint values h_{min}, h_{max} and volume Vol. The recursive iteration-scheme employed for the numerical solution of the minimization problem is based on the following observation. For regions in the plate of intermediate thickness, $h_{min} < h < h_{max}$, the first condition in (42) can be written as:

$$(\frac{\partial D_{\alpha\beta\kappa\gamma}}{\partial h} e_{\alpha\beta} e_{\kappa\gamma})(\Lambda(1 - \mu)(1 - \gamma))^{-1} = 1. \tag{45}$$

From this, a recursion formula for the thickness can be constructed: We write A_k for the left hand side of (45) at the k'th iteration step and then

$$h_{k+1} = \begin{cases} \max\{1 - \zeta)h_k, h_{min}\} & \text{if } h_k A_k^n < \max\{(1 - \zeta)h_k, h_{min}\} \\ h_k A_k^n & \text{if } \max\{(1 - \zeta)h_k, h_{min}\} < h_k A_k^n < \min\{(1 + \zeta)h_k, h_{max}\} \\ \min\{(1 + \zeta)h_k, h_{max}\} & \text{if } h_k A_k^n \geqslant \min\{(1 + \zeta)h_k, h_{max}\} \end{cases} \tag{46}$$

for a given number n and a maximum stepsize ζ. Similar recursion formulas can be constructed from (42b), (42c) for the densities μ and γ. The iterative scheme is then as follows:

Assign starting point (h, μ, γ, θ) and values ζ and n.

I. For values $(h_k, \mu_k, \gamma_k, \theta_k)$ of thickness, densities and direction of stiffeners, compute the deflection w_k by a finite difference method. Compute the compliance Π_k. If $\Pi_k > \Pi_{k-1}$, increase the step size ζ and restart at step $(k - 1)$.

II. By numerical differentiation compute the curvature tensor e_k and the derivatives of $D_{\alpha\beta\kappa\gamma}^k$ with respect to h, μ, γ. Compute θ_{k+1} by solving (42d) for fixed h_k, μ_k, γ_k (combined Newton-bisection method).

III. Find Λ_k so that $h_{k+1}, \mu_{k+1}, \gamma_{k+1}$ given by recursion formulas of the form (46) satisfy the volume constraint (this gives an inner iteration loop, where the volume constraint equation is solved by a bisection method). The new design is $(h_{k+1}, \mu_{k+1}, \gamma_{k+1}, \theta_{k+1})$.

IV. If a convergence criterion has not been satisfied, go to I.

The values of ζ and η have to be found by experiment. It was found that best results are obtained if the thickness and densities can be assigned different values of the maximum step size. The convergence criterion was set on the compliance

The necessary conditions (42)-(44) can also be treated algebraically and such an analysis gives for example that a purely solid sub-region of intermediate thickness will never appear at a simply supported or free edge of an optimally designed plate; this behaviour is also seen from the computations.

Some numerical results obtained for rectangular, clamped plates are shown in Table 1 below, and Fig. 7 shows the configuration of an optimal plate.

The optimal designs and the optimal values of the objective function for the extended design space are very stable with respect to the discretisation used, see Figs. 7 and 8. The numerical experiments thus indicate that the inclusion of ribstiffened plates in the set of admissible designs regularizes the optimization problem. For axisymmetric plates this result has been obtained by Cheng and Olhoff ([7]) and Olhoff et. al ([17]); for that type of plates allowing for one field of stiffeners apparently regularizes the problem.

$h_{min} = 0.2;\ h_u = 0.35;\ \nu = 0.25$	$h_{max} = 0.5$	$h_{max} = 1.0$
With two fields of stiffeners modelled as plate elements	0.468	0.156
With one field of stiffeners modelled as plate elements	0.477	0.164
With two fields of stiffeners modelled as beam elements	0.475	0.174
With only the slowly varying thickness as design variables	0.498	0.224

Table 1. Values of Π/Π_u for optimal design of a uniformly loaded rectangular clamped plate (sides with ratio of lengths 3 :2). Volume is fixed and corresponds to a plate of uniform thickness h_u with compliance Π_u (the reference design). The optimal plate has compliance Π. For the unstiffened plate the optimal value shown is obtained using a 30 $\times$ 30 mesh.

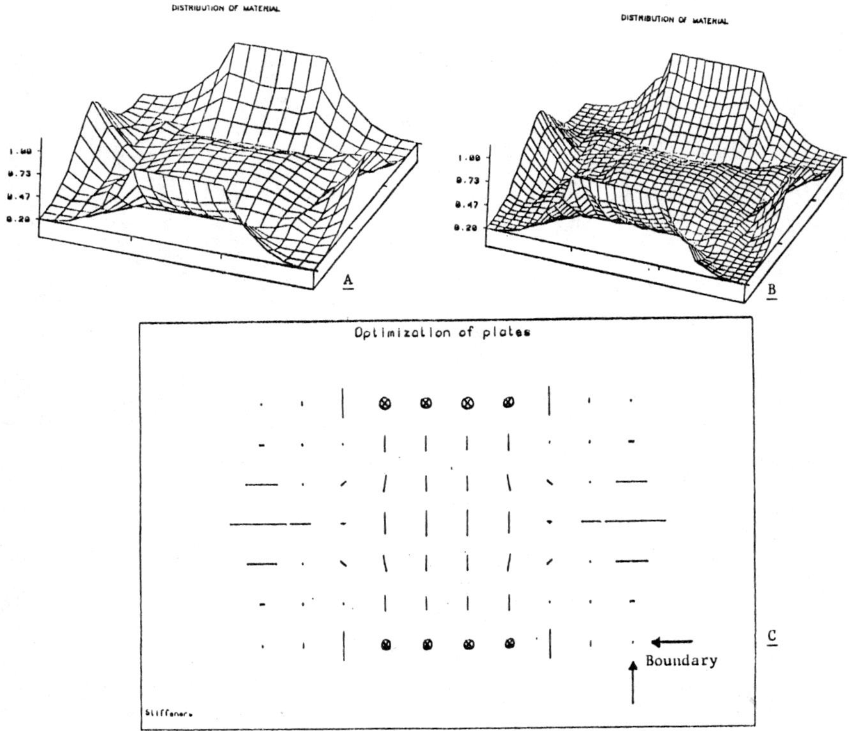

Fig. 7. Optimal design of a rib-stiffened, clamped plate. $\underline{A}$ and $\underline{B}$ shows the distribution of material over the plate which in this case implies the distribution of material in the stiffeners, as the solid part of the plate has thickness $h = h_{max}$ (distribution is the pointwise volume of material $= (1-\mu)(1-\gamma)h + (\mu+\gamma-\mu\gamma)h_{max}$); $\underline{A}$: for a 17×17 mesh, $\underline{B}$: for a 30×30 mesh. $\underline{C}$ shows the directions and densities of the stiffeners; the length of each line-segment shows the density of stiffeners in direction of that line-segment. $\otimes$ indicates that one density is equal to one in that point, so the plate is solid with thickness h_{max}. (Values of constraints are $h_{min} = 0.2$, $h_{max} = 1.0$ and $h_u = 0.35$).

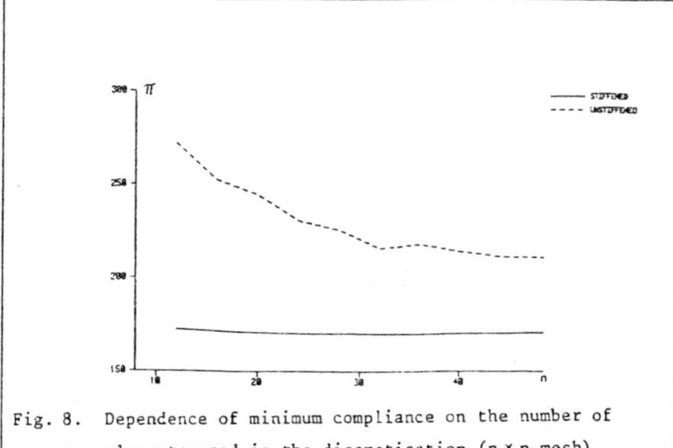

Fig. 8. Dependence of minimum compliance on the number of elements used in the discretisation ($n \times n$ mesh).

From table 1 we see that the rib stiffened plates are superior to the nonstiffened ones. We also see that, as expected, modelling the stiffeners as plate elements (rigidity tensor given by (38)) gives stiffer plates than in the case where the ribs are modelled as beams (rigidity tensor given by (39)). However the difference between the two types of modelling is not great, from which we conclude that our method of smear-out employed to obtain the effective rigidity tensor for the case of plate-element stiffeners does not result in an grossly overestimated stiffness. It should be noted, however, that when optimizing the beam rib-reinforced plates, regions of the plate with densities of stiffeners equal to one ($\mu = 1$ or $\gamma = 1$) were treated as being regions of solid plate with thickness $h = h_{max}$ and not as a region completely occupied by beams with no interaction. We also note that very efficient designs can be obtained by use of just one field of stiffeners. The configuration of stiffeners obtained in our continuous setting gives a very good indication of how to construct a stiff plate with just a finite number of stiffeners (Fig. 7c; see also Ref. [7]). This is of course of great practical importance. Also of practical importance is the fact that optimal designs within the extended design space can be obtained by use of a very coarse discretisation; a similar phenomenon has been observed for other regularized design problems (Kohn-private communication).

The optimization described above treats plates reinforced with integral ribs, but the analysis and results would be very similar for the optimization of composite plates reinforced with fibers in the solid part of the plate. Our results suggest that considerable improvement in the performance of a fiber reinforced plate could be obtained by an optimization using both the densities as well as the directions of the fibers as design variables.

VI. Conclusion

The results reported here confirm that optimization of rib stiffened plates seems to be a well-posed problem, not only for axisymmetric plates as studied elsewhere, but also in the general two-dimensional case. Also, the rib stiffened plates give rise to very efficient designs. The study thus gives good reason to

believe that the G-closure problem for the fourth order plate operator can be
solved using the method of laminates first used by Lurie et. al. for various
second order problems. However, very little is known about deriving G-limits of
fourth order operators, so there is plenty of work to be done.

REFERENCES

[1] M. Artola, G. Duvaut: Homogénéisation d'une classe de problemes non
lineares. C.R. Acad. Sc. Paris, t. 288 (1979), Serie A, pp. 775-778.

[2] M.P. Bendsøe: Some smear-out models for integrally stiffened plates with
applications to optimal design. Proc. Int. Symp. Optimum Structural Design,
Oct. 1981, Univ. Ariz., Tucson, Arizona. Pineridge Press Ltd., Swansea 1982,
pp. 13-29 to 13-34.

[3] M.P. Bendsøe: On obtaining a solution to optimization problems for solid,
elastic plates by restriction of the design space. J. Struct. Mech., Vol. 11
(1983), pp. 501-521.

[4] M.P. Bendsøe: Existence proofs for a class of plate optimization problems.
Proc. 11th IFIP Conf. on System Modelling and Optimization. Copenhagen 1983.
Lecture Notes in Control and Information Sciences, Vol. 59 (1984), Springer
Verlag, pp. 773-779.

[5] K.T. Cheng, N. Olhoff: An Investigation Concerning Optimal Design of Solid
Elastic Plates. Int. Solids Struct., Vol. 17 (1981), pp. 305-323.

[6] K.T. Cheng: On Non-Smoothness in Optimal Design of Solid, Elastic Plates.
Int. J. Solids Struct., Vol. 17 (1981), pp. 795-810.

[7] K.T. Cheng, N. Olhoff: Regularized Formulation for Optimal Design of
Axisymmetric Plates. Int. J. Solids Struct., Vol. 18 (1982), pp. 153-169.

[8] G. Duvaut: Comportement macroscopique d'une plaque perforée periodiquement.
Lecture Notes in Mathematics, Vol. 594, Springer, Berlin, 1977, pp. 131-145.

[9] E.J. Haug, B. Rousselet: Design Sensitivity Analysis in Structural
Mechanics, I. Static Response Variations, J. Struct. Mech., Vol. 8 (1980),
pp. 17-41.

[10] R.V. Kohn, M. Vogelius: A New Model for Thin Plates with Rapidly Varying
Thickness. Int. J. Solids Struct., Vol. 20, (1984), pp. 333-350.

[11] K.A. Lurie, A.V. Cherkaev, A.V. Fedorov: Regularization of Optimal Design
Problems for Bars and Plates. JOTA, Vol. 37 (1982), pp. 499-522 (Part I), and
pp. 523-543 (Part II).

[12] K.A. Lurie, A.V. Cherkaev, A.V. Fedorov: On the Existence of Solutions to
Some Problems of Optimal Design for Bars and Plates. JOTA, Vol. 42 (1984),
pp. 247-282.

[13] K.A. Lurie, A.V. Cherkaev: G-closure of some Particular Sets of Admissible
Material Characteristics for the Problem of Bending of Thin Plates. JOTA,
Vol. 42 (1984), pp. 305-316.

[14] F.I. Niordson: Optimal Design of Plates with a Constraint on the Slope of
     the Thickness Function. Int. J. Solids Struct., Vol. 19 (1983), pp. 141-151.

[15] N. Olhoff: Optimal Design of Vibrating Circular Plates. Int. J. Solids
     Struct., Vol. 6 (1970), pp. 139-156.

[16] N. Olhoff: Optimal Design of Vibrating Rectangular Plates. Int. J. Solids
     Struct., Vol. 10 (1974), pp. 93-109.

[17] N. Olhoff, K.A. Lurie, A.V. Cherkaev, A.V. Fedorov: Sliding Regimes and
     Anisotropy in Optimal Design of Vibrating Axisymmetric Plates. Int. J. Solids
     Struct., Vol. 17 (1981), pp. 931-948.

[18] G.I.N. Rozvany, N. Olhoff, K.-T. Cheng, J.E. Taylor: On the Solid Plate
     Paradox in Structural Optimization. J. Struct. Mech., Vol. 10 (1982), pp.
     1-32.

[19] E. Sanchez-Palencia: Non-Homogeneous Media and Vibration Theory. Lecture
     Notes in Physics, Vol. 127, Springer Verlag, 1980.

<u>Note</u> <u>added</u> <u>in</u> <u>proof</u>:

Recent work by Gibiansky and Cherkaev ([A]) shows that laminates of laminates
(of second rank) can indeed be used to construct optimal microstructures for plates. This implies that for large values of h_{max}/h_{min} and h_{max}/h_u , where an optimal plate has either $h = h_{min}$ or $h = h_{max}$ (bang-bang type thickness function),
the plate with minimum compliance can be realised by use of a stiffener-like
structure which corresponds to a laminate of laminates. This result has been
proved by employing the method of compensated compactness (quasiconvexification)
as a tool to obtain estimates for the effective moduli of plate-composites; these
estimates then turn out to be exact as the bounds can be achieved by certain laminates of laminates of second rank.

[A]: L.V. Gibiansky, A.V. Cherkaev: Design of Composite Plates of Extremal
      Rigidity. Report No. 914 of the A.F. Ioffe Physical Technical Institute,
      Academy of Sciences of the USSR, Leningrad 1984. (in Russian).

THE EFFECTIVE DIELECTRIC COEFFICIENT OF A COMPOSITE MEDIUM:
RIGOROUS BOUNDS FROM ANALYTIC PROPERTIES

by

David J. Bergman*
School of Physics and Astronomy
Tel-Aviv University
Tel-Aviv 69978   ISRAEL

and

Corporate Research Science Laboratories
Exxon Research and Engineering Co.
Route 22 East
Annandale, NJ   08801

<u>Abstract</u>

The analytic properties of the bulk effective dielectric coefficient ε_e of a composite medium, viewed as a function of the component coefficients $\varepsilon_1, \varepsilon_2, \ldots$, are reviewed in Section I, and are then used to discuss rigorous bounds for (real and complex) ε_e when various types of partial information are available about the medium. General methods are described in Section II for constructing optimal and rigorous bounds for both real and complex ε_e in two-component composites. A method for constructing rigorous bounds for ε_e in composites made of more than two components is described in Section III. Optimum bounds are found for real ε_e. For complex ε_e the optimization problem has not been fully resolved. Nevertheless, bounds for complex ε_e are obtained that are better than some other bounds that have recently been derived in Refs. 6 and 7.

---

* permanent address

I. Introduction and summary of the known properties of ε_e

Consider a sample of a composite material made of $n+1$ pure isotropic components $\varepsilon_1 \ldots \varepsilon_{n+1}$, that completely fills the volume of a large parallel plate condenser. When appropriate boundary conditions are applied (i.e., a fixed potential at each plate and zero normal field at the side walls), the electric field inside the condenser $\vec{E}(\vec{r})$ is usually determined but has a complicated dependence on position due to the microscopic inhomogeneity of the filler medium. The bulk effective dielectric coefficient ε_e is defined by comparing the electrostatic energy of the actual filler to that of a fictitious homogeneous filler, where the field would be a constant $\vec{E}_0$ everywhere

$$\varepsilon_e E_0^2 \equiv \frac{1}{V} \int dV \ \varepsilon(\vec{r}) \ E^2(\vec{r}) \ . \qquad (I-1)$$

Here $\varepsilon(\vec{r})$ is the local value of the dielectric coefficient, which can be written in terms of the ε_i and the characteristic (step) functions $\theta_i(\vec{r})$ as follows

$$\varepsilon(\vec{r}) = \sum_{i=1}^{n+1} \varepsilon_i \theta_i(\vec{r}) \ , \qquad (I-2)$$

and V is the volume of the condenser. We note that the composite sample does not have to be isotropic or cubic: In the anisotropic case our definition leads to a linear combination of the principal values of the tensor ε_e that is selected by the direction of $\vec{E}_0$ relative to the principal axes.

The analytical properties of the function $\varepsilon_e(\varepsilon_1,\varepsilon_2,\ldots\varepsilon_{n+1})$ can be discussed in terms of the following quantities

$$\nabla\phi(\vec{r}) \equiv \vec{E}(\vec{r})/|\vec{E}_0|$$

$$h_i \equiv \frac{\varepsilon_i}{\varepsilon_{n+1}}, \ i = 1\ldots n$$

$$h_{n+1} \equiv 1 \qquad (I-3)$$

$$m(h_1,h_2\ldots h_n) \equiv \frac{\varepsilon_e}{\varepsilon_{n+1}}$$

$$\theta_h(\vec{r}) \equiv \sum_{i=1}^{n+1} h_i \ \theta_i(\vec{r}) \ ,$$

which take advantage of the fact that ε_e as well as $\vec{E}(\vec{r})$ are order one homogeneous functions of $\varepsilon_1 \ldots \varepsilon_{n+1}$. The potential field $\phi(\vec{r})$ satisfies a partial differential equation with boundary conditions on the surface of the condenser

$$\left\{ \begin{array}{l} \nabla \cdot (\theta_h \nabla \phi) = 0 \\ \qquad \phi = 0, L \quad \text{for} \quad z = 0, L \quad \text{(on the plates)} \\ \dfrac{\partial \phi}{\partial n} = 0 \quad \text{on the side walls} , \end{array} \right\} \qquad (I\text{-}4)$$

where the plates are taken to be perpendicular to the Z-axis, and the distance between them is L. In terms of these quantities (I-1) becomes

$$m(h_1 \ldots h_n) = \frac{1}{V} \int dV \, \theta_h (\nabla \phi)^2 . \qquad (I\text{-}5)$$

Two alternative forms of this result can be obtained by first transforming (I-5) to an integral over the surface of the condenser

$$\int dV \theta_h (\nabla \phi)^2 = \int dV [\nabla \cdot (\phi \theta_h \nabla \phi) - \phi \nabla \cdot (\theta_h \nabla \phi)]$$
$$= \oint \phi \theta_h \frac{\partial \phi}{\partial n} \, dA , \qquad (I\text{-}6)$$

and then noting that, due to the boundary conditions, ϕ can be replaced in the surface integral either by $\phi_0 \equiv z$ or by the complex conjugate of the potential ϕ^* (this is relevant only when some of the h_i's, and hence also ϕ , are complex). The inverse transformation then leads to the two expressions

$$m(h_1 \ldots h_n) = \frac{1}{V} \int dV \theta_h \frac{\partial \phi}{\partial z} = \frac{1}{V} \int dV \theta_h |\nabla \phi|^2 . \qquad (I\text{-}7)$$

It is now easy to prove that, in any (complex) neighborhood of $h \equiv (h_1 \ldots h_n)$ where (I-4) has a unique solution, $m(h)$ is analytic.[1,2] Any singular point of $m(h)$ must therefore be a point where a non-trivial solution $\psi(\vec{r})$ exists for the zero boundary value counterpart of (I-4). Applying the same transformation that was used in (I-6) we can show that in this case

$$\int dV \theta_h |\nabla \psi|^2 = \oint dA \, \psi^* \, \theta_h \frac{\partial \psi}{\partial n} = 0 , \qquad (I\text{-}8)$$

where the surface integral vanishes because of the zero boundary conditions on ψ. Eqs. (I-7) and (I-8) have these important consequences for the analytic properties of m(h):

(a) $m(h^*) = [m(h)]^*$

(b) Whenever all the h_i are real and positive, m(h) is regular, real and positive.

(c) Whenever all the values of $Im(h_i)$ are nonzero and have the same sign, m(h) is analytic and $Im(m(h))$ also has the same sign.

(d) When all $h_i = 1$, then $\phi \equiv \phi_0 \equiv z$ and hence $m(1,1,...) = 1$.

(e) In the case of a two-component composite, m(h) depends on only one variable. From (a) and (b) it then follows that m can be singular only if h is real and negative, and that it can be represented as follows[3]

$$m(h) = 1 - F(s) \tag{I-9a}$$

where

$$s \equiv \frac{1}{1-h} \tag{I-9b}$$

$$F(s) = \int_0^1 \frac{\mu(x)dx}{s-x} , \tag{I-9c}$$

and where $\mu(x)dx$ is a non-negative measure that satisfies

$$\mu(x) = \pi \ Im[F(x-i0)] = \pi \ Im \ [m(1 - \frac{1}{x} + i0)]. \tag{I-10}$$

Further properties of $m(h)$ are obtained by considering values of h_i that are near 1. In this way we can show that

(f) $$\left(\frac{\partial m}{\partial h_i} \right)_{\text{all } h_j = 1} = p_i = \text{the volume fraction of } \epsilon_i \tag{I-11}$$

(g) $$\left(\frac{\partial^2 m}{\partial h_i^2} \right)_{\text{all } h_j = 1} = -\frac{1}{d} p_i(1-p_i) \quad \text{and} \tag{I-12}$$

$$\left(\frac{\partial^2 m}{\partial h_i \partial h_j} \right)_{\text{all } h_k = 1} = \frac{1}{d} p_i p_j \quad \text{for } i \neq j$$

in the case of a composite with isotropic or cubic symmetry under rotations.

Here d is the dimensionality (in practice, d is either 2 or 3).

In the two component case (i.e., $n=1$), representations such as (I-9) lead to a multitude of exact bounds,[1,2,4,5,19-21] as I shall demonstrate in section II. When $n>1$ attempts to construct representations which are appropriate generalizations of (I-9) have been made, and these have also been used to derive bounds.[6,7] Here I will pursue a different approach for this case, which sometimes leads to better bounds. This will be discussed in section III. Eq. (I-9) can also form the basis of a detailed calculation of ε_e starting from accurate knowledge of the microgeometry of the composite mixture. This will not be described here – the reader is referred to Refs. 2,5,8-10 for a discussion of this approach as well as for some practical examples. I will also not discuss some consequences for the characteristic singularity that appears in $\varepsilon_e(\varepsilon_1,\varepsilon_2)$ when one of the components is at some kind of connectivity threshold – these matters are discussed in Refs. 1,5,11,12.

II. Rigorous bounds for ε_e in a two-component composite

These are conveniently discussed in terms of the following functions of the variable s

$$s \equiv \frac{1}{1-h} = \frac{\varepsilon_2}{\varepsilon_2 - \varepsilon_1}$$

$$F(s) \equiv 1 - m(h) = 1 - \frac{\varepsilon_e}{\varepsilon_2} \tag{II-1}$$

$$E(s) \equiv 1 - \frac{h}{m(h)} = \frac{1-sF(s)}{s(1-F(s))} = 1 - \frac{\varepsilon_1}{\varepsilon_e}.$$

Both of these functions are analytic in the entire complex s-plane except for the real segment [0,1], and they both have representations similar to (I-9), e.g.

$$F(s) = \int_0^1 \frac{\mu(x)dx}{s-x} \tag{II-2}$$

where $\mu(x)dx$ is a positive measure. The rigorous bounds which can be constructed for $F(s)$ or $E(s)$ from various types of partial information about the composite always involve a measure that is concentrated at a finite number of points. We shall anticipate this by representing $F(s)$ (and $E(s)$) as a discrete sum of simple poles with positive residues

$$F(s) = \sum \frac{F_n}{s - f_n} \quad , \quad 0 < f_n < 1 \quad , \quad 0 < F_n \quad ;$$

$$\tag{II-3a}$$

$$E(s) = \sum \frac{E_n}{s - e_n} \quad , \quad 0 < e_n < 1 \quad , \quad 0 < E_n \quad .$$

A further important property of these functions is

$$F(1) < 1 \; , \; E(1) < 1. \tag{II-3b}$$

In fact, I shall define a class Z of functions by the properties (II-3a), (II-3b). All functions E and F defined as in (II-1) with reference to a finite dielectric composite with a sufficiently smooth interface between components, can be shown to belong to this class.[2] This class is closely related to the class of Stieltjes functions whose bounding properties are discussed in Ref. 14. We now discuss some further properties of these functions and methods for bounding them.

First of all we note that the results (I-11), (I-12) for the derivatives of $m(h)$ become sum rules for the zeroth and first moments of the pole spectra of F and E, namely

$$\sum F_n = \sum E_n = p_1$$
$$\sum f_n F_n = \frac{1}{d} \, p_1 p_2 \tag{II-4a}$$
$$\sum e_n E_n = \frac{d-1}{d} \, p_1 p_2$$

where the last two sum rules hold for an isotropic or cubic composite, and where $p_1, p_2 = 1-p_1$ are the volume fractions of the two components. Next, we note that the relationship between F and E can be expressed as follows

$$(1-F(s))(1-E(s)) = \frac{s-1}{s} \; . \tag{II-4b}$$

From this it is clear that the infinities (i.e., poles) of one function occur at values of s for which the other function is equal to 1, except for s=0 or 1. Other singularities (e.g., cuts, branch points, essential singularities) must occur simultaneously in both functions. Finally, at s = 0 if one function is regular the other has a pole, while at s=1 if one is regular and less than 1, then the other function is regular and equal to 1.

For the case of real s outside the semi-closed segment $[0,1)$, a basic inequality for F(s) follows from (II-3), namely

$$F(S) > \frac{1}{s} \sum_n F_n = \frac{p_1}{s} .$$ (II-5a)

If we substitute for F and s in terms of ε_e, ε_1, ε_2 we obtain

$$\varepsilon_e < p_1 \varepsilon_1 + p_2 \varepsilon_2 ,$$ (II-5b)

i.e., a well known bound for real ε_e (one of the so-called Wiener bounds)[22]. Applying the same reasoning to E(s) we find

$$E(s) > \frac{1}{s} \sum_n E_n = \frac{p_1}{s} ,$$ (II-6a)

and hence

$$\frac{1}{\varepsilon_e} < \frac{p_1}{\varepsilon_1} + \frac{p_2}{\varepsilon_2} ,$$ (II-6b)

which is the other Wiener bound. Note that the explicit value of the bound depends on the zero moment of the pole spectrum, which in turn is known if the volume fractions are known.

For the case of s that lies anywhere else in the complex plane, we write the following basic inequality which again follows from the representation (II-3),

$$\mathrm{ReF}(s) + \frac{\mathrm{Res}}{\mathrm{Ims}} \mathrm{ImF}(s) = \frac{\mathrm{Im}(sF(s))}{\mathrm{Ims}} = - \sum_n \frac{f_n F_n}{|s-f_n|^2} < 0. $$ (II.7)

This inequality constrains F(s) to lie on one side of a straight line in the complex F-plane. A similar inequality for E(s) constrains it to lie on one side of a straight line in the E-plane. When this latter constraint is transformed to the F-plane using (II-1), we find that F must now lie on the inside of a circle (since E(s) and F(s) for fixed s are obtained from each other by means of a Möbius transformation). Note that in obtaining these complex bounds no information about any moments of the pole spectra were used. These bounds, which must both be satisfied simultaneously, are easily translated into bounds for complex ϵ_e.[20,23]

In order to obtain better bounds, more information about the system must be used, e.g., the values of the zeroth and first moments of the pole spectrum given in (II-4a). In fact, higher moments of the pole spectrum can also be calculated if more detailed information is available concerning the microgeometry in the form of correlation functions such as $\langle \theta_1(\vec{r}_1)\theta_1(\vec{r}_2)\dots \rangle$, where the angular brackets signify averaging over the volume (see, e.g., Refs. 5 and 13).

The simplest way to introduce such information into the bounding process is to define a hierarchy of functions of the class Z that incorporate increasing amounts of such information in their definition. Thus, we define

$$C^{(0)}(s) \equiv F(s)$$

$$C^{(1)}(s) \equiv \frac{1 - (\Sigma\, F_n)/sF(s)}{1 - \Sigma\, F_n}$$

$$\vdots$$

$$C^{(r+1)}_{(s)} \equiv \frac{1 - (\Sigma C^{(r)}_n)/sC^{(r)}_{(s)}}{1 - \Sigma\, C^{(r)}_{(n)}}$$

(II-8)

where each function in the hierarchy needs the zeroth moment of its predecessor for its definition. It is straightforward to show that the function $C^{(r)}(s)$ is of class Z, and that the k-th moment of its pole spectrum depends on the moments of $C^{(r-1)}(s)$ of order 0 to k+1. E.g., in an isotropic system, for which both the zeroth and the first moment of F are known, we find

$$C^{(1)}(s) = \frac{1}{p_2} - \frac{p_1}{p_2 sF(s)} = \sum_n \frac{C^{(1)}_n}{s - s_n}$$

$$\sum_n C^{(1)}_n = \frac{\Sigma f_n F_n}{(\Sigma F_n)(1 - \Sigma F_n)} = \frac{1}{d} \; .$$

(II-9)

Moments of $C^{(1)}$ higher than the zeroth moment are not known, while the information about the zeroth moment of F has been used to ensure that $C^{(1)}(s)$ belongs to class Z. A similar hierarchy of functions $D^{(r)}(s)$ can be constructed starting from $D^{(0)}(s) = E(s)$. From the definitions of $C^{(r)}(s)$ and $D^{(r)}(s)$ we can show that they are related to each other in the same way as $F(s)$ and $E(s)$, namely

$$(1-C^{(r)}(s))(1-D^{(r)}(s)) = \frac{s-1}{s} \qquad (II-10a)$$

The proof is by induction on the order r, and proceeds by first showing from (II-10a) that the sum of the zero moments of $C^{(r)}$ and $D^{(r)}$ is unity

$$\sum_n C_n^{(r)} + \sum_n D_n^{(r)} = 1, \qquad (II-10b)$$

and then using the recursive definition of $C^{(r+1)}$ and $D^{(r+1)}$ to calculate the product $(1-C^{(r+1)})(1-D^{(r+1)})$.

An improved bound on $F(s)$ for real $s \notin [0,1)$ can now be obtained by applying the procedure of (II-5a) to $C^{(1)}(s)$, namely

$$C^{(1)}(s) > \frac{1}{s} \sum_n C_n^{(1)} = \frac{1}{sd} . \qquad (II-11a)$$

This is easily translated into a bound for $F(s)$, and finally to a bound for real ε_e, namely

$$\varepsilon_e < \varepsilon_2 + \frac{p_1}{(\varepsilon_1-\varepsilon_2)^{-1} + p_2/d\varepsilon_2} \quad \text{for } \varepsilon_1 < \varepsilon_2 . \qquad (II-11b)$$

This is one of the Hashin-Shtrikman bounds[15] - the other one is obtained by repeating the entire procedure for $D^{(1)}(s)$.

An improved bound for complex $F(s)$ is obtained from the same function $C^{(1)}(s)$ by applying to it the procedure of (II-7). This leads to a straight line bound for $C^{(1)}$ and hence to a circle bound for $F(s)$ and ε_e.

A similar procedure which starts from $D^{(1)}(s)$ leads to another circle bound for $F(s)$ and ε_e.[20,23] Thus $F(s)$ must lie within the intersection of two circles. Note that in contrast to the real bounds (II-11) obtained from $C^{(1)}$, the complex bounds do not use the information from the first moment of $F(s)$. Hence they are not restricted to isotropic or cubic composites. In order to use this first moment to get yet better complex bounds we must advance another step in the (II-8) hieracrchy and consider the following class Z function.

$$C^{(2)}(s) = \frac{1}{d-1} \left(d - \frac{p_2 F(s)}{sF(s) - p_1} \right). \tag{II-12}$$

When the procedure of (II-7) is applied to this, an improved circular bound is obtained for $F(s)$. As before, another improved circular bound is obtained by starting from $D^{(2)}$ instead of $D^{(2)}$.

These bounds for complex $F(s)$, first derived independently by Bergman[19,23] and Milton[20,31], are described in greater detail in Ref. 4, together with explicit representations for the bounding lines. The question of optimality is also discussed: In all cases the boundary lines (or points in the case of real F) are very simple class Z functions, namely, only a small number of poles appear in the expansion of type (II-3) (see, e.g., the bound of (II-5a)). In many cases these simple functions correspond to ε_e for an exactly solvable microgeometry, and then we know that the bounds are achievable and hence optimal. However, the functions F that are derived from ε_e of a physical composite must obey some equalities[24] or inequalities[25] that result from the following "phase exchange relationships" satisfied by $\varepsilon_e(\varepsilon_1, \varepsilon_2)$

$$\varepsilon_e(\varepsilon_1, \varepsilon_2) \cdot \varepsilon_e(\varepsilon_2, \varepsilon_1) = \varepsilon_1 \cdot \varepsilon_2 \qquad \text{for} \quad d = 2$$

$$\varepsilon_e(\varepsilon_1, \varepsilon_2) \cdot \varepsilon_e(\varepsilon_2, \varepsilon_1) > \varepsilon_1 \cdot \varepsilon_2 \qquad \text{for} \quad d \neq 2 \text{ when } \varepsilon_1, \varepsilon_2 > 0. \tag{II-13}$$

Therefore these F-functions form a more restricted class than the entire class Z. In some cases the bounds found for general class Z functions violate the restrictions resulting from (II-13) and are thus certainly unachievable. By explicitly including those added restrictions, one can then improve upon the bounds derived for general class Z functions. This was first done for the 2D case by Milton[20,31] and for the 3D case by Bergman[4,5]. In pactice, however, the improvement is rather modest[4,5] and we will therefore not pursue this subject any further here.

Another type of information that can be used to obtain improved bounds are known values of ε_e at different values of $\varepsilon_1, \varepsilon_2$ but for the same microgeometry.[26] In practice such information can be obtained by measuring ε_e in the same sample at a different frequency or temperature, if ε_1 or ε_2 change with those parameters. Such information can also be obtained by measuring a different physical property, such as thermal conductivity, electrical conductivity, magnetic permeability, diffusivity, all of which are described by the same mathematical function. E.g., the bulk effective thermal conductivity $\kappa_e(\kappa_1, \kappa_2)$ is the same function as $\varepsilon_e(\varepsilon_1, \varepsilon_2)$ for a given sample. Such information is translated into known values of $F(s_1)$, $F(s_2)$, etc. This information can now be used to construct improved bounds for $F(s)$ at any s. These bounds are optimal within the class Z. We note that the possibility of obtaining such detailed information about $F(s)$ from a knowledge of $F(s_1)$ would not exist were $F(s)$ just an arbitrary analytic function. It is entirely due to the close connection between the class Z and the class of Stieltjes functions[14]. Bounds for real $F(s)$ or $\varepsilon_e(\varepsilon_1, \varepsilon_2)$ which incorporate one known real value of $F(s_1)$ were first obtained by Prager,[26] and subsequently improved by Bergman.[27,28] The general problem of bounds for real or complex $F(s)$ when any number of (real or complex) values of $F(s_1)$, $F(s_2)$, etc. are known was first discussed by Milton.[21] Milton and Golden[29] subsequently pointed out the connection between this problem and the best error bounds for Padé approximants discussed by Baker.[14,30] I will present these general bounds in a different way - in the form of a set of definitions and theorems (without proof) that can easily be utilized to produce the bounds in practice.

<u>Definition 1</u>:

$$Z_M = \left\{ F(s): F(s) = \sum_1^N \frac{F_n}{s-f_n} \text{ if } M = 2N; \; F(s) = \sum_1^N \frac{F_n}{s-f_n} + \frac{F_{N+1}}{s} \right.$$
$$\left. \text{if } M = 2N + 1 \; ; \; F_n, f_n = \text{real} \; ; \; 0 < f_n < 1 \; ; \right.$$
$$\left. 0 < F_n \qquad ; \qquad F(1) < 1 \right\}$$

<u>Remark</u>: Clearly, the function classes Z_M form a nested sequence culminating in Z

$$Z_1 \subset Z_2 \subset Z_3 \ \cdots \cdot \subset Z.$$

<u>Definition</u> 2: A set of pairs of (real or complex) numbers $(s_i, F^{(i)})$ $i = 1 \ldots M$ is called "compatible" iff $\exists F(s) \in Z$ such that $F(s_i) = F^{(i)}$, and it is called "N-compatible" iff $\exists F(s) \in Z_N$ such that $F(s_i) = F^{(i)}$.

<u>Remark</u>:

Clearly N-compatibility implies compatibility. That the converse is also sometimes true is highly non-trivial. This is a consequence of the following theorems. Bounds for the case when all the relevant quantities are real result from the following two theorems.

<u>Theorem</u> 1: Given a set of M compatible real pairs $(s_i, F^{(i)})$ such that $s_i \notin [0,1)$, then that set is also M-compatible. Moreover, there exists a unique $F_a(s) \in Z_M$ such that $F_a(s_i) = F^{(i)}$. $F_a(s)$ is continuous in $F^{(i)}$.

<u>Remark</u>: Applying the transformation $F(s) \to E(s)$ we get $(s_i, F^{(i)}) \to (s_i, E^{(i)})$ and using Theorem 1 a unique $E_b(s) \in Z_M$ such that $E_b(s_i) = E^{(i)}$. This can be transformed back to a unique $F_b(s) \in Z_{M+1}$ such that $F_b(s_i) = F^{(i)}$, $F_b(1) = 1$, and $F_b(s)$ is continuous in $F^{(i)}$.

<u>Theorem</u> 2: (bounds for the totally real case) If we enlarge the (compatible real) set $(s_i, F^{(i)})$, $s_i \notin [0,1)$, by adding one more real pair $(s_{M+1}, F^{(M+1)})$ such that $s_{M+1} \notin [0,1)$, then the combined set is compatible iff $F^{(M+1)}$ lies between $F_a(s_{M+1})$ and $F_b(s_{M+1})$.

Bounds for the case when all the relevant quantities have nonzero imaginary parts result from the following two theorems.

<u>Theorem</u> 3: A set of M compatible complex pairs (s_i, F^i) is also $2M$-compatible, and this corresponds to a unique class Z_{2M} function if all pairs have a nonzero Im or $\mathrm{Im} F^{(i)}$. Furthermore, there exists a unique set of functions $F_a(s|\lambda) \in Z_{2M+1}$, where λ is a continuous real parameter $\lambda_1 < \lambda < \lambda_2$, such that $F_a(s_i|\lambda) = F^{(i)}$ for all $\lambda \in [\lambda_1, \lambda_2]$. $F_a(s|\lambda)$ is continuous in $F^{(i)}$ and λ.

Remark: The same procedure applied to $(s_i, E^{(i)})$ yields another function $F_b(s|\mu) \in Z_{2M+2}$ such that $F_b(s_i|\mu) = F^{(i)}$ and $F_b(1|\mu) = 1$ for all $\mu \in [\mu_1, \mu_2]$. For fixed s, $F_a(s|\lambda)$ and $F_b(s|\mu)$ are parametric represetnations of two circular arcs which meet at their extremities (i.e., when $\lambda = \lambda_1$ or λ_2 and $\mu = \mu_1$ or μ_2) and define a compact region in the complex F-plane - the intersection of two circles - denoted by $R[s]$.

Theorem 4 (bounds for the totally complex case): If we add one more pair $(s_{M+1}, F^{(M+1)})$ to the existing complex set, then the combined set is compatible iff $F^{(M+1)}$ lies inside $R[s_{M+1}]$.

Remarks: (a) Other cases, where part of the information is real and other parts are complex, can be treated by obvious extensions of the above theorems.
(b) When information is available also about moments of the pole spectrum, we must first select the appropriate $C^{(r)}(s)$ function from the hierarchy of (II-8). Using the above theorems we find bounds for $C^{(r)}$, and those are then translated back to bounds for $F(s)$ or ϵ_e. In the complex case the bound will invariably be given by the intersection of two circles (or a circle and a straight line) in the complex plane.

An example of a real bound resulting from one known real value $F(s_1)$ can be found in Ref.1, while the improved bound that results for this case when the inequality of (II-13) is also included is derived in Ref. 5. The case of a complex bound resulting from one known complex value $F(s_1)$ has not been worked out before, so we now proceed to derive it. We first seek a function $F_a(s)$ of the form

$$F_a(s) = \frac{F_0}{s - f_0} + \frac{F_1}{s} \tag{II-14a}$$

such that

$$F_a(s_1) = F(s_1). \tag{II-14b}$$

This function will depend on one free parameter, which we could choose to be F_1. In that case, F_1 will range from 0 up to a maximum value F_{1max} determined by the requirement $F_a(s|F_{1max}) = 1$. Those two values correspond to the extremities of a circular arc and are common to the other circular arc, the form of which is

$$E_b(s) = \frac{E_0}{s-e_0} + \frac{E_1}{s} \rightarrow F_b(s) = \frac{F_0'}{s-f_0'} + \frac{F_1'}{s-f_1'} \tag{II-15}$$

The form exhibited for $F_b(s)$ is easily derived by considering the transformation $E(s) \rightarrow F(s)$. In practice, it is more convenient to parametrize $F_a(s)$ by f_0, while using the fact that $F_1 = 0$ determines one of the common points of the two arcs $F_a(s|f_0)$, $F_b(s|e_0)$. We denote this point by A. The other common point, which we denote by B, is determined by the requirement $E_1 = 0$ applied to $E_b(s|e_0)$. The explicit solution of (II-14) yields

$$F_0(f_0) = - \frac{|s_1-f_0|^2 \, \mathrm{Im}(s_1 F(s_1))}{f_0 \, \mathrm{Im} s_1}$$

$$F_1(f_0) = \frac{1}{\mathrm{Im}(\frac{1}{s_1})} \left(\mathrm{Im} F(s_1) - \frac{1}{f_0} \mathrm{Im}(s_1 F(s_1)) \right). \tag{II-16}$$

By setting $F_1(f_0) = 0$ we can now find the position of the point A in the complex F-plane

$$A_F = \frac{|F(s_1)|^2 \, \mathrm{Im} s_1}{\mathrm{Im}(s_1 F(s_1) - s \mathrm{Im} F(s_1)} \tag{II-17}$$

By analogy, the point B has a similar representation in the complex E-plane, namely

$$B_E = \frac{|E(s_1)|^2 \, \mathrm{Im} \, s_1}{\mathrm{Im}(s_1 E(s_1)) - s \mathrm{Im} \, E(s_1)} . \tag{II-18}$$

Because we know that the arc described by (II-16) is a part of a circle, an alternate way to describe it is by giving one more point on that circle, e.g., the point determined by $f_0 \rightarrow \infty$. If we denote this point by C, then its position in the F-plane is given by

$$C_F = \frac{1}{\mathrm{Im} s_1} \left(\mathrm{Im}(s_1 F(s_1)) - \frac{|s_1|^2}{s} \mathrm{Im} \, F(s_1) \right). \tag{II-19}$$

The other arc is part of another circle that is represented in the E-plane by expressions for $E_0(e_0)$, $E_1(e_0)$ that are the same as (II-16), except that $E(s_1)$

appears instead of $F(s_1)$. Alternatively, we can characterize the circle by the points A,B and a third point D, which corresponds to $e_0 \to \infty$. The position of this point in the E-plane is given by an expression which is the analogue of (II-19), namely

$$D_E = \frac{1}{\mathrm{Im}\, s_1} \left(\mathrm{Im}(s_1 E(s_1)) - \frac{|s_1|^2}{s} \mathrm{Im}E(s_1) \right). \qquad (\mathrm{II}\text{-}20)$$

III. Rigorous bounds for ε_e in a composite with more than two components

The analytic properties of $m(h_1 \ldots h_n)$ for $n > 1$ are more difficult to manipulate than in the two component case $n=1$. Although some progress has recently been made in constructing generalizations of the representation (I-9),[6,7,16,32] I will discuss a different approach here. An important aspect of the dielectric propeties of a multicomponent composite is that the same result must be obtained for ε_e when we replace the pure components $\varepsilon_1 \ldots \varepsilon_{n+1}$ by a set of two-component composites with the same values for their bulk effective dielectric coefficients $\varepsilon_1(\varepsilon_1', \varepsilon_2') \ldots \varepsilon_{n+1}(\varepsilon_1', \varepsilon_2')$, if each of these two-component composites has a sufficiently fine microstructure. Since the scale of the microstructure can always be made arbitrarily small without altering the function $\varepsilon_i(\varepsilon_1', \varepsilon_2')$, this does not constitute any limitation. (In this connection, see the discussion of the effects of a dilation transformation given at the end of Section 2 of Ref. 9.) The consequence of this is that if we substitute for h_i any function $h_i = m_i(h)$ that describes some two-component composite, then the resulting composite function $m(h) = m(m_1(h) \ldots m_n(h))$ also describes a two component composite. I will assume without proof that this property continues to hold even when $m_i(h)$ and $m(h)$ are allowed to range over the slightly larger class of functions that satisfy the properties (a) - (e) from Section I and can be represented in the form (I-9).

We can now apply to $m(h)$ all the procedures discussed in the previous section for obtaining bounds. Since there is considerable freedom in choosing the trajectories $h_i(h)$, we must try and choose them so as to make the resulting bounds as narrow as possible. As we shall see, sometimes this optimization of the trajectories can be carried out and the best possible bound is then obtained. In other cases, we do not yet know how to choose optimal trajectories. We can then still construct bounds by choosing particular trajectories, though these do not necessarily yield the best possible bounds. In what follows I will discuss three-component systems. The generalization to any number of components greater than two is immediate.

From the representation (I-9), it is a straightforward matter to calculate the derivatives of $m(h)$

$$m'(h) = s^2 \int_0^1 \frac{\mu(x)dx}{(s-x)^2}$$

$$m''(h) = -2s^3 \int_0^1 \frac{\mu(x)xdx}{(s-x)^3}$$

$$s \equiv (1-h)^{-1} .$$

(III-1)

Thus, $m'(h) > 0$ for any real h, while $m''(h) < 0$ for any real and positive h. Consequently, since $m(1) = 1$, the following inequality is satisfied for real h

$$1 - m(h) \geq m'(1)(1-h) \quad \text{for} \quad h \geq 0. \tag{III-2}$$

(Alternatively, one could note that this is just a rewritten form of the inequality (II-5a).)

In order to apply this inequality to the composite function $m(h_1(h),h_2(h))$ we must calculate its derivative at $h=1$. Recalling that $h_1(1) = h_2(1) = m(1,1) = 1$ and using the results for the partial derivatives of m from (I-11), we find

$$\frac{d}{dh} m(h_1(h),h_2(h))\big|_{h=1} = p_1 h_1'(1) + p_2 h_2'(1) , \tag{III-3}$$

and consequently, from (III-2),

$$1-m(h_1(h),h_2(h)) \geq (p_1 h_1'(1) + p_2 h_2'(1))(1-h) \quad \text{for} \quad h \geq 0 . \tag{III-4}$$

This constitutes a bound for real $m(h_1,h_2)$ that depends, however, on the choice of trajectory $h_1(h),h_2(h)$. The best of these bounds would be obtained if we could maximize the r.h.s. of (III-4) by allowing $h_1(h)$ and $h_2(h)$ to vary over all allowable functions that lead to a trajectory that passes through the desired point (h_1,h_2). In this case this is easily achieved by invoking the inequality (III-2) for each of the functions $h_1(h),h_2(h)$

$$1-h_i(h) > h_i'(1)(1-h) \quad \text{for} \quad h > 0. \tag{III-5}$$

Using this to maximize the r.h.s. of (III.4) we find

$$1-m(h_1(h),h_2(h)) > p_1(1-h_1(h)) + p_2 (1-h_1(h)), \tag{III-6}$$

a result which does not depend on the precise form of the trajectory. Therefore we can ignore the dependence on h and rewrite the bound in the trajectory independent form

$$1 - m(h_1,h_2) > p_1(1-h_1) + p_2(1-h_2) \quad \text{for} \quad h_1,h_2 > 0 , \tag{III-7a}$$

which when translated back to ε_e reads

$$\varepsilon_e(\varepsilon_1,\varepsilon_2,\varepsilon_3) < p_1\varepsilon_1 + p_2\varepsilon_2 + p_3\varepsilon_3 \quad \text{for all} \quad \varepsilon_i > 0. \tag{III-7b}$$

In order to find another bound for $m(h_1,h_2)$, we consider a different but related function $\tilde{m}(\tilde{h}_1,\tilde{h}_2)$

$$\tilde{h}_i \equiv \frac{1}{h_i}$$

$$\tilde{m}(\tilde{h}_1,\tilde{h}_2) \equiv \frac{1}{m(h_1,h_2)} \tag{III-8}$$

which can be shown to have the same properties as $m(h_1,h_2)$.[1] We now choose a trajectory in the $\tilde{h}_1,\tilde{h}_2$ space of the form $\tilde{h}_1 = \tilde{m}_1(\tilde{h})$, $\tilde{h}_2 = \tilde{m}_2(\tilde{h})$, and the entire discussion which led to (III-7) can then be repeated. The final results are

$$1-\tilde{m}(\tilde{h}_1,\tilde{h}_2) > p_1(1-\tilde{h}_1) + p_2(1-\tilde{h}_2)$$

$$\frac{1}{\varepsilon_e(\varepsilon_1,\varepsilon_2,\varepsilon_3)} < \frac{p_1}{\varepsilon_1} + \frac{p_2}{\varepsilon_2} + \frac{p_3}{\varepsilon_3} . \tag{III-9}$$

In order to include more information about the rotational symmetry of the composite, we must consider second derivatives of $m(h_1,h_2)$ and also of $h_1(h),h_2(h)$. Though possible, this would seriously complicate the problem of optimizing the trajectory - Ref. 1 uses that approach. Here we adopt a different strategy, namely, we define a new function of new variables, related to $m(h_1,h_2)$, but with second derivatives that vanish when $h_i = 1$: (this transformation is related to one that was used in Eq. (8.40) of Ref. 6 and Eq. (5.16) of Ref. 32)

$$\upsilon_i \equiv \frac{1}{\frac{1}{1-h_i} - \frac{1}{d}} \quad , \quad i=1,2$$

$$q(\upsilon_1,\upsilon_2) \equiv \frac{1}{\frac{1}{1-m(h_1,h_2)} - \frac{1}{d}} \cdot \qquad \text{(III-10)}$$

The variables υ_i are similar to $1-h_i$, while q is similar to $1-m$. More specifically, if we consider the two-component case, then $q(\upsilon)$ can be singular only for h real and negative, i.e., for υ that is real and lies outside the segment $(-d,\frac{d}{d-1}]$. Restricting ourselves to the case of a discrete spectrum of poles, it is easy to show that the poles of $q(\upsilon)$ are all simple and have negative residues. We can then represent $q(\upsilon)$ as follows

$$q(\upsilon) = A_0\upsilon + \sum_n \frac{A_n\upsilon}{\upsilon_n(\upsilon_n-\upsilon)} ; \qquad \text{(III-11)}$$

$$A_0, A_n > 0 \quad \text{and} \quad \upsilon_n < -d < \quad \text{or} \quad \upsilon_n > \frac{d}{d-1} \ .$$

It follows that

$$q'(\upsilon) = A_0 + \sum_n \frac{A_n}{(\upsilon-\upsilon_n)^2} > 0 \quad \text{for real} \quad \upsilon$$

$$q''(\upsilon) = - \sum_n \frac{2A_n}{(\upsilon-\upsilon_n)^3} \quad \begin{matrix} > \\ < \end{matrix} \ 0 \qquad " \qquad \text{(III-12)}$$

$$q'''(\upsilon) = \sum_n \frac{6A_n}{(\upsilon - \upsilon_n)^4} > 0 \qquad " \ \ .$$

For the two-component as well as for the multicomponent case we find, from (III-10)

$$q(\text{all } \upsilon_i = 0) = 0$$

$$dq(\text{all } \upsilon_i = 0) = \sum_i p_i d\upsilon_i \qquad \text{(III-13)}$$

$$d^2q(\text{all } \upsilon_i = 0) = 0 \ .$$

The last of these equalities is valid for an isotropic or cubic composite. Finally, we note that in the two-component case, since both $q(\upsilon)$ and $q''(\upsilon)$ are monotonic for real υ and vanish at $\upsilon=0$, they must have a constant sign for positive (negative) υ within the "physical region" $(-d,\frac{d}{d-1})$, i.e.

$$g(\upsilon),g''(\upsilon) < 0 \quad \text{for} \quad -d < \upsilon < 0$$
$$g(\upsilon),g''(\upsilon) > 0 \quad \text{for} \quad 0 < \upsilon < \frac{d}{d-1} \, . \tag{III-14}$$

Thus the following inequalities hold

$$g(\upsilon) \lessgtr g'(0)\upsilon < 0 \quad \text{for} \quad \upsilon \in (-d,0)$$

$$\tag{III-15}$$

$$g(\upsilon) \gtrless g'(0)\upsilon > 0 \quad \text{for} \quad \upsilon \in (0,\frac{d}{d-1}) \, .$$

As before, we invoke the idea that if in a three-component composite we substitute for υ_1 and υ_2 a pair of two-component composites characterized by the g-functions $\upsilon_i = g_i(\upsilon)$, then the resulting composite function $g(g_1(\upsilon),g_2(\upsilon))$ is again a valid two-component g-function. Applying (III-15) to this function we find

$$g(\upsilon_1(\upsilon),\upsilon_2(\upsilon)) \lessgtr (p_1\upsilon_1'(0) + p_2\upsilon_2'(0)) \, \upsilon \quad \text{for} \quad \begin{Bmatrix} \upsilon \in (-d,0) \\ \upsilon \in (0,\frac{d}{d-1}) \end{Bmatrix} . \tag{III-16}$$

The r.h.s. of this inequality depends on the choice of the trajectory $\upsilon_1(\upsilon)$, $\upsilon_2(\upsilon)$, but we can optimize the bound by using (III-15) once again - this time for $\upsilon_1(\upsilon)$ and $\upsilon_2(\upsilon)$. In this way we obtain, in analogy with (III-7), an inequality that is independent of any trajectory

$$g(\upsilon_1,\upsilon_2) \lessgtr p_1\upsilon_1 + p_2\upsilon_2 \quad \text{for} \quad \upsilon_1,\upsilon_2 \in \begin{Bmatrix} (-d,0) \\ (0,\frac{d}{d-1}) \end{Bmatrix} . \tag{III-17}$$

When these inequalities are translated back to ε_e language, they are found to be just the Hashin-Shtrikman bounds for a three-component composite.[15]

In constrast to the real case, the problem of finding bounds for complex

$\varepsilon_e(\varepsilon_1,\varepsilon_2,\varepsilon_3)$ from similar information has not yet been solved in a satisfactory way, except for the simplest case, i.e., when no information is available about the microgeometry and only the (complex) values of $\varepsilon_1,\varepsilon_2,\varepsilon_3$ are known. In the complex ε_e-plane one draws the bounds for the three two-component composites $(\varepsilon_1,\varepsilon_2)$, $(\varepsilon_2,\varepsilon_3)$, $(\varepsilon_3,\varepsilon_1)$. The bound for, e.g., $(\varepsilon_1,\varepsilon_2)$ is the intersection of the half plane, defined by the straight line through $\varepsilon_1,\varepsilon_2$, which does not include the origin 0, and the inside of the circle through $\varepsilon_1,\varepsilon_2$, and 0.[4,19,23] The straight line segments from the edges of the three two-component bounds define a triangle with linear edges and with vertices at $\varepsilon_1,\varepsilon_2,\varepsilon_3$. Similarly, the circular arcs from those edges define a triangular region with the same vertices but with the arcs as its three edges. The union of the two triangular regions is the allowed region in the complex plane where $\varepsilon_e(\varepsilon_1,\varepsilon_2,\varepsilon_3)$ must lie[6,18].

An attempt to obtain improved bounds for the complex case by including information about the volume fractions and about rotational symmetry has recently been made[6,7] which employs a representation for $F(s_1,s_2)$ that is a generalization of (II-2). The bounds are however not rigorous because some unproved assumptions were used regarding the form of the extreme points of a set of two-dimensional positive measures.[6,7] I will now discuss the alternative approach to this problem, which is based on considering a trajectory $1/s_i = F_i(s)$, as was done before for the real case. This always leads to rigorous, if not optimal, bounds. Applying this method to a specific example, I will obtain a better bound than was obtained in Ref. 6 for the same example.

We look for a trajectory in the form of a single pole expression for each of the $1/s_i$

$$\frac{1}{s_i} = \frac{a_i}{s - b_i/a_i} \, , \quad a_i > b_i > 0, \, a_i^2 < a_i - b_i \, , \qquad \text{(III-18)}$$

where the inequalities must be satisfied in order to ensure that $1/s_i(s)$ belongs to the class Z of two-component functions. Writing the complex variables s,s_1,s_2 in terms of their real and imaginary parts

$$s = \sigma + i\rho \, , \quad s_i = \sigma_i + i\rho_i \qquad \text{(III-19)}$$

the coefficients a_i, b_i will be given by

$$a_i = \frac{\rho}{\rho_i} \quad , \quad b_i = \frac{\rho\sigma}{\rho_i} - \frac{\rho^2 \sigma_i}{\rho_i^2} \quad , \tag{III-20}$$

while the inequalities can be reformulated as

$$\frac{\rho}{\rho_i} > 0 \; ; \; \frac{\rho\sigma}{\rho_i} > \frac{\rho^2 \sigma_i}{\rho_i} \; ; \; \frac{\rho}{\rho_i}(1-\sigma) > \frac{\rho^2}{\rho_i^2}(1-\sigma_i) \; . \tag{III-21a}$$

A necessary consequence of the first of these inequalities is

$$0 < \frac{\rho_1}{\rho_2} \; , \tag{III-21b}$$

which may not be satisfied by s_1, s_2. It is, however, always possible to rename the dielectric coefficients $\varepsilon_1, \varepsilon_2, \varepsilon_3$ in such a way (in fact, in exactly two different ways) that (III-21b) is satisfied. (For an n-component composite, (III-21b) generalizes to the requirement that all $\mathrm{Im} s_i$ have the same sign, and again this can be achieved in exactly two different ways by renaming ε_i.) Obviously, the sign of ρ must be the same as that of ρ_i . We thus obtain the following restrictions on ρ and σ :

$$0 < \sigma - \frac{\sigma_i}{\rho_i}\rho \; ; \; \sigma + \rho\,\frac{1-\sigma_i}{\rho_i} < 1. \tag{III-22}$$

Choosing a value for $s = \sigma + i\rho$ within these bounds, we try to find bounds for the class Z function $F(s_1(s), s_2(s))$ where $s_i(s)$ are given by (III-18). The asymptotic behavior of this function for large $|s|$ is

$$F(s_1(s), s_2(s)) \longrightarrow \frac{p_1}{s_1(s)} + \frac{p_2}{s_2(s)} + \frac{a_1 p_1 + a_2 p_2}{s} \; , \tag{III-23}$$

as $|s| \to \infty$, and therefore we can bound it by the following two circular arcs[4]

$$F_c(s) = \frac{a_1 p_1 + a_2 p_2}{s - s_0} \quad , \; 0 < s_0 < 1 - a_1 p_1 - a_2 p_2$$

$$F_d(s) = \frac{1}{s}\left(1 - \frac{1 - a_1 p_1 - a_2 p_2}{s_0}\right) + \frac{1 - s_0}{s - s_0}\,\frac{1 - a_1 p_1 - a_2 p_2}{s_0} \; , \tag{III-24}$$

$$1 - a_1 p_1 - a_2 p_2 < s_0 < 1 \; .$$

The two arcs meet at their two extremities, namely, at the points

$$B = \frac{a_1 p_1 + a_2 p_2}{s} \; , \quad C = \frac{a_1 p_1 + a_2 p_2}{s-1 + a_1 p_1 + a_2 p_2} \; . \tag{III-25}$$

The circle represented by $F_c(s)$ also passes through the origin 0 , while the circle represented by $F_d(s)$ also passes through the point $1/s$. Thus the points $B, C, 0, 1/s$ determine the two circles, and hence the bounds, completely.

As an example, we consider the following composite

$$\varepsilon_1 = -4 + 4i \qquad \varepsilon_2 = i \qquad \varepsilon_3 = 4 + 4i$$
$$p_1 = 0.45 \qquad p_2 = 0.1 \qquad p_3 = 0.45 \tag{III-26a}$$

which leads to

$$s_1 = \frac{1+i}{2} \qquad s_2 = \frac{4}{25}(7+i) \tag{III-26b}$$

$$\rho > 0 \; , \quad \sigma - 7\rho > 0 \; , \quad \sigma + \rho < 1.$$

Choosing s so as to satisfy the last two inequalities as equalities, namely

$$s = \frac{7+i}{8} \; , \tag{III-26c}$$

we find

$$B = \frac{61}{2000}\,(7-i) \; ; \; C = \frac{61}{2041}\,(21-40i) \; ; \; \frac{1}{s} = \frac{4}{25}(7-i), \tag{III-26d}$$

which, together with the origin 0, define two circular bounds F_c, F_d for $F \equiv 1 - \varepsilon_e/\varepsilon_3$. Renaming the dielectric coefficients as follows

$$\varepsilon_1' = i \qquad \varepsilon_2' = 4 + 4i \qquad \varepsilon_3' = -4 + 4i \tag{III-27a}$$

$$p_1 = 0.1 \qquad p_2 = 0.45 \qquad p_2 = 0.45$$

we find

$$s_1' = \frac{4}{25} (7-i) \qquad s_2' = \frac{1-i}{2}$$

$$\rho' < 0 \quad , \quad \sigma' + 7\rho' > 0 \ , \ \sigma' - \rho' < 1. \tag{III-27b}$$

Choosing s' again so as to satisfy the last two inequalities as equalities, namely

$$s' = \frac{7-i}{8} \ , \tag{III-27c}$$

we find two circular bounds F_c', F_d' for $F' \equiv 1 - \varepsilon_e/\varepsilon_3'$ defined by the origin $0'$ and the points

$$B' = \frac{61}{2000} (7+i) \quad ; \quad C' = \frac{61}{2041} (21 + 40i) \ ; \ \frac{1}{s'} = \frac{4}{25} (7+i) \tag{III-27d}$$

These bounds, as well as the previous pair described in (III.26), are easily translated to the complex ε_e - plane, and are displayed in Fig. 1, along with the bounds obtained for the same example in Ref. 6 using a different method. The fact that the bounds (III-26) , (III-27) are somewhat better than those of Ref. 6, even though obtained in a more rigorous fashion and without attempting to optimize the choice of trajectories, demonstrates that the method of choosing trajectories is a useful one. Moreover, since this manuscript was completed, I have been able to show that these bounds have certain optimality properties.[17]

References

1. D.J. Bergman, Physics Reports 43, 377-407 (1978), also published as Willis E. Lamb, Jr. - a festschrift on the occasion of his 65-th birthday, eds. D. ter Haar and M.O. Scully, (North Holland, 1978), pp. 377-407.

2. D.J. Bergman, in Les Méthodes de l'Homogénéisation: Théorie et Applications en Physique, Ecole d'Été d'Analyse Numérique, Editions Eyrolles, Paris, 1985.

3. A different derivation of Eq. (I-9) based on a stochastic model for the composite medium has been given by K. Golden and G. Papanicolaou, Comm. Math. Phys. 90, 473 (1983).

4. D.J. Bergman, Annals of Physics 138, 78-114 (1982)

5. D.J. Bergman, in Macroscopic Properties of Disordered Media, Eds. R. Burridge, S. Childress, and G. Papanicolaou, (Springer-Verlag 1982) pp. 10-37.

6. K.M. Golden, Ph.D. Thesis, Courant Institute for Mathematical Sciences, (unpublished)

7. K. Golden and G. Papanicolaou, To appear in J. Stat. Phys.

8. D.J. Bergman, J. Phys. $\underline{C12}$, 4947-4960 (1979).

9. D.J. Bergman, Phys. Rev. $\underline{B19}$, 2359-2368 (1979)

10. Y. Kantor and D.J. Bergman, J. Phys. $\underline{C15}$, 2033-2042 (1982)

11. D.J. Bergman and Y. Imry, Phys. Rev. Letters $\underline{39}$, 1222-1225 (1977)

12. D.J. Bergman, AIP Conf. Proc. No. $\underline{40}$, pp. 46-61 (1978), Eds. J.C. Garland and D.B. Tanner

13. G.W. Milton, Phys. Rev. Letters $\underline{46}$, 542 (1981)

14. G.A. Baker, Jr., J. Math. Phys. $\underline{10}$, 814 (1969)

15. Z. Hashin and S. Shtrikman, J. Appl. Phys. $\underline{33}$, 3125-3131 (1962).

16. G.F. Dell'Antonio, R. Figari, and E. Orlandi, Unpublished

17. D.J. Bergman, and G. Milton, to be published.

18. G. Milton, private communication.

19. D.J. Bergman, Phys. Rev. Lett. $\underline{44}$, 1285 (1980)

20. G.W. Milton, Appl. Phys. Lett. $\underline{37}$, 300 (1980)

21. G.W. Milton, J. Appl. Phys. $\underline{52}$, 5294-5304 (1981)

22. O. Wiener, Abh. Sachs. Akad. Wiss. Leipzig Math.-Naturwiss. Kl. $\underline{32}$, 509 (1912)

23. D.J. Bergman, Phys. Rev. $\underline{B23}$, 3058 (1981)

24. J.B. Keller, J. Math. Phys. $\underline{5}$, 548 (1964).

25. K. Schulgasser, J. Math. Phys. $\underline{17}$, 378 (1976).

26. S. Prager, J. Chem. Phys. $\underline{50}$, 4305 (1969).

27. D.J. Bergman, Phys. Rev. $\underline{B14}$, 1531-1542 (1976).

28. D.J. Bergman, Phys. Rev. $\underline{B14}$, 4304-4312 (1976).

29. G.W. Milton and K. Golden, preprint.

30. G.A. Baker, Jr., "Essentials of Padé approximants", Academic Press, New York, 1975.

31. G.W. Milton, J. Appl. Phys. $\underline{52}$, 5286-5293 (1981)

32. K. Golden, preprint submitted to J. Mech. Phys. Solids.

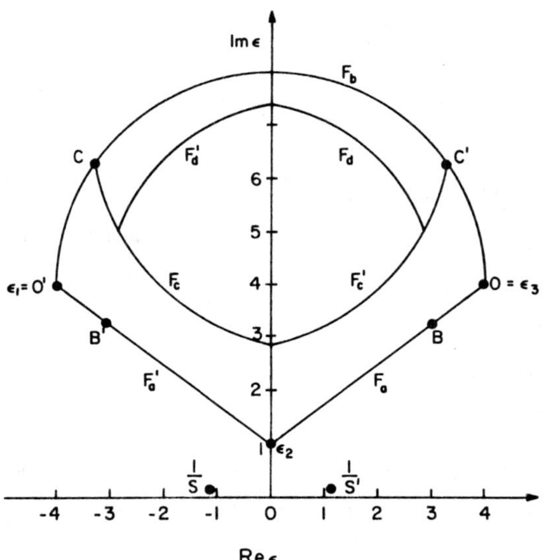

<u>Figure 1</u>:   Exact bounds on complex ε_e for the example discussed in the text: $\varepsilon_1 = -4 + 4i$, $\varepsilon_2 = i$, $\varepsilon_3 = 4 + 4i$; $p_1 = 0.45$, $p_2 = 0.1$, $p_3 = 0.45$. The straight lines F_a , F'_a and the semi-circle F_b define the bound which is obtained when the information about p_i is ignored. The circular arcs F_c , F_d , F'_c , F'_d are the edges of the improved bound found in this work when the information about p_i is included. For the same example, Ref. 6 obtains as an improved bound the area whose edges are F_c, F'_c , F_b . The arcs F_c, F'_c can also be obtained by optimizing the chosen trajectories.

VARIATIONAL BOUNDS ON DARCY'S CONSTANT

James G. Berryman
Lawrence Livermore National Laboratory
P.O. Box 808, L-201
Livermore, CA 94550

Abstract

Prager's variational method of obtaining upper bounds on the fluid permeability (Darcy's constant) for slow flow through porous media is reexamined. By
exploiting the freedom one has in choosing the trial stress distributions, several
new results are derived. One result is a phase interchange relation for permeability; when the fluid-phase and particle-phase are interchanged for a fixed
geometry, we find an upper bound on a linear combination of the complementary permeabilities. Another result is a proof of the monotone properties of the bounds.
The optimal two-point bounds from this class of variational principles are evaluated numerically and compared to exact results of low density expansions for
assemblages of spheres.

1. Introduction

Darcy's law [1] states that, when a viscous fluid moves slowly and steadily
through a porous medium, the macroscopic flux (fluid volume crossing a unit cross-
sectional area per unit time) is directly proportional to the pressure difference
across the material and inversely proportional both to its thickness and to the
viscosity of the fluid. The direction of flow is opposite to that of the positive
pressure difference. The constant of proportionality in Darcy's law is called the
fluid permeability or Darcy's constant. The law is modified slightly if gravity
influences the fluid motion as it did in Darcy's original experiments; then the
driving force is not simply the pressure difference but a linear combination of
the forces due to gravity and pressure differential. For simplicity, gravita-
tional effects will be neglected in the present discussion.

Darcy's law is a linear proportionality between the macroscopic flux and
pressure gradient. This macroscopic relationship can be derived from the

microscopic (Stokes) equations for slow flow of an incompressible fluid through a
vessel of arbitrary shape with no slip between fluid and vessel at points of contact. (The terms "macroscopic" and "microscopic" are being used somewhat loosely
here to distinguish between the two relevant length scales.) Derivations of
Darcy's law using averaging methods have been given by Poreh and Elata [2] and by
Neuman [3] while homogenization theory was used in the derivation by Keller [4].
The feature of this problem which makes it both especially interesting and especially difficult is the fact that, although only classical physics is involved at
both the macroscopic and microscopic length scales, the macroscopic law of
transport is not of the same form as the microscopic flow equations. The permeability is therefore a meaningful macroscopic concept without a direct microscopic analog. By contrast, the dielectric constant in Maxwell's equations is
meaningful both at the microscopic level and at the macroscopic level for composites. Only when we consider much smaller (i.e., atomic) length scales does a
similar change in the _form_ of the relevant equations occur.

One early and fairly successful attempt to estimate the macroscopic permeability for flow through a random porous aggregate is the work of Brinkman [5].
By postulating a modified form of Darcy's law which might be expected to apply
when the flow field is nonuniform, Brinkman was able to perform a single-site
scattering calculation for flow around a spherical inclusion in a porous medium
and then to use this result to obtain an effective medium estimate of the permeability as a function of porosity. For very low densities of solid particles,
Brinkman's approach gives the leading corrections to the Stokes drag on a spherical particle in the presence of other particles [6-8]. More recently some exact
calculations of the permeability for periodic arrays of spheres have been performed by Zick and Homsy [9] and by Sangani and Acrivos [10]. All of these efforts
are important benchmarks in the general theory of permeability but they fall short
of helping us to estimate the permeability of an arbitrary porous random aggregate
of particles.

Prager [11] introduced a very general variational principle for porous flow
based on the concept of minimum energy dissipation. Upper bounds on permeability

may be obtained from this minimum principle if certain statistical information concerning the topology of the pore space is available. Doi [12] introduced another variational principle for bounds on permeability but this approach does not have the flexibility inherent to Prager's method. Torquato [13] was the first to apply Prager's ideas to packings of penetrable and impenetrable spheres for which the required statistical correlation functions are known for large densities of particles. The present author repeated some of these calculations independently [14] and subsequently discovered that some different choices of trial stress distribution led to significantly different upper bounds on Darcy's constant [15]. The hope of obtaining realistic estimates of permeability when the required spatial correlation functions of an arbitrary porous specimen are known [16] remains unfulfilled at present but the progress which has been made will be reported here.

Section 2 presents the variational principle which leads to bounds on permeability. Section 3 explores the possible choices of trial stress distribution. Section 4 uses the freedom in choice of stochastic function in the trial stress distribution to derive a phase interchange relation for permeability. Section 5 discusses optimal upper bounds on permeability using only two-point spatial correlation functions which have been derived using the freedom in choice of deterministic function in the trial stress distribution. Section 6 illustrates the monotone properties of the bounds and summarizes our conclusions.

2. Minimum Energy Dissipation

In this section, we will present a complete derivation of the variational principle to be used in the remainder of the paper. The initial line of argument parallels that of the elastic problem as presented by Courant and Hilbert [17]. The conclusion of the derivation has been alluded to by Prager [11] and by Beran [18]; however, since some questions have been raised concerning a choice of normalization in the formulation [19,20], it will prove beneficial to include a full discussion here.

Equations for the Stokes flow of a viscous fluid through an arbitrary vessel of total volume Ω take the form

$$\Pi_{ij,j} = 0 \qquad \text{for} \quad i, j = 1, 2, 3 , \tag{2.1}$$

where the stress tensor is

$$\Pi_{ij} = -p\delta_{ij} + \sigma_{ij} , \tag{2.2}$$

the local fluid pressure is p, and the viscosity stress tensor is

$$\sigma_{ij} = \mu(v_{i,j} + v_{j,i}) \tag{2.3}$$

with μ the viscosity of the fluid. A subscript following a comma indicates a partial derivative. The local fluid velocity is v_i and if the fluid is assumed to be incompressible

$$v_{i,i} = 0 . \tag{2.4}$$

The summation convention is assumed in both (2.1) and (2.4). The boundary conditions associated with (2.1) and (2.4) are the no slip condition

$$v_i = 0 \quad \text{on} \quad \Gamma_I \tag{2.5}$$

at interior points where the fluid touches the vessel and the stress matching condition

$$\Pi_{ij}n_j = \Pi^{o}_{ij}n_j \quad \text{on} \quad \Gamma_E \tag{2.6}$$

at points on the external boundary of the volume Ω . The fluid volume being considered is completely contained in the volume Ω_f whose surface is $\Gamma = \Gamma_I \cup \Gamma_E$. If the applied stress $\Pi_{ij}n_j$ is not uniformly hydrostatic on Γ_E , then fluid will flow into Ω_f through parts of the surface Γ_E and out of Ω_f through other parts of Γ_E .

For simplicity, we will generally assume a slab geometry. The vessel is then a porous material lying between two parallel planes orthogonal to the z-axis. The

56

applied stress takes the form of a uniform pressure p_- on one plane located at $z = -\Delta Z/2$ and $p_+ = p_- + \Delta P$ on the other at $z = \Delta Z/2$. If the thickness of the slab is ΔZ , then the pressure gradient is $\frac{\Delta P}{\Delta Z}$. The total volume Ω for the slab is infinite. The discussion which follows is phrased as if Ω were finite with the understanding that the limit $\Omega \to \infty$ will be taken at the end of the calculations.

Now we wish to reformulate (2.1) - (2.6) in terms of a variational principle. To do so, we introduce the quadratic form

$$Q(\sigma,\sigma) = \frac{1}{2\mu\Omega} \int_{\Omega_f} \sigma_{ij}\sigma_{ij} \ d^3x \ . \tag{2.7}$$

Eq. (2.7) gives the rate of energy dissipation in the fluid per unit total volume [21]. Now if θ and τ are two symmetric, zero trace tensors and if θ is a viscosity stress tensor for some fluid velocity u_i , then

$$\theta_{ij} = \mu \ (u_{i,j} + u_{j,i}) \ \text{in} \ \Omega_f \tag{2.8}$$

and

$$Q(\theta,\tau) = \frac{1}{2\mu\Omega} \int_{\Omega_f} \theta_{ij} \ \tau_{ij} \ d^3x$$

$$= -\frac{1}{\Omega} \int_{\Omega_f} u_i \ \tau_{ij,j} \ d^3x + \frac{1}{\Omega} \int_{\partial\Omega_f} u_i \tau_{ij} n_j ds \ . \tag{2.9}$$

Eq. (2.9) is a Green identity which follows easily from an application of the divergence theorem; the infinitesimal surface element is ds and n_j is the j-th component of the unit outward normal.

Now define two comparison tensors θ^o and τ^o which satisfy the following conditions:

$$\theta_{ij}^o = \mu(u_{i,j}^o + u_{j,i}^o) \ \text{in} \ \Omega_f \tag{2.10}$$

where u^o satisfies $u_{i,i}^o = 0$ (because of the zero trace property of θ^o) and

$$u_i^o = 0 \ \text{on} \ \Gamma_I \ , \tag{2.11}$$

while

$$\tau^{o}_{ij,\,j} = p^{o}_{,i} \quad \text{in } \Omega_f \tag{2.12}$$

and

$$\tau^{o}_{ij} n_j = (\Pi^{o}_{ij} + p^{o}\,\delta_{ij})n_j \quad \text{on } \Gamma_E . \tag{2.13}$$

Consider symmetric, zero trace tensors θ and τ such that θ satisfies (2.10) and (2.11) for some velocity field u_i, not necessarily u^{o}_i, and that, like τ^{o}, τ also satisfies equation (2.13) and another of the form (2.12) for some scalar p (instead of p^{o}). Then it follows from (2.9) that

$$Q(\theta - \theta^{o}, \tau - \tau^{o}) = 0 \tag{2.14}$$

for all such θ and τ. Thus, the tensors $\theta - \theta^{o}$ and $\tau - \tau^{o}$ are orthogonal with respect to Q.

With these definitions, we may now formulate two reciprocal [17] variational problems

$$Q(\theta - \tau^{o}, \theta - \tau^{o}) = \text{minimum} \tag{2.15}$$

and

$$Q(\tau - \theta^{o}, \tau - \theta^{o}) = \text{minimum}. \tag{2.16}$$

After eliminating known quantities, these two principles reduce to

$$Q(\theta,\theta) - \frac{2}{\Omega} \int_{\Gamma_E} u_i \Pi^{o}_{ij} n_j \, ds = \text{minimum} \tag{2.17}$$

and

$$Q(\tau,\tau) = \text{minimum}, \tag{2.18}$$

subject to the admissibility conditions (2.10) - (2.11) and (2.12) - (2.13) respectively. The absolute minimum for both (2.17) and (2.18) is achieved by the symmetric tensor $\sigma = \theta = \tau$ where σ satisfies (2.1) - (2.6).

Furthermore, it is worth remarking that the admissibility conditions for one principle are the variational (or Euler) equations for the other.

The first variational principle (2.17) is associated with the name of Helmholtz [11], [22]. This principle is difficult to apply in problems with random geometry because of the no slip condition (2.11) on the trial stress fields θ. A trivial but nevertheless admissible choice of trial velocity field is $u_i = 0$ everywhere. This choice places an upper bound of zero on the right hand side of (2.17). The general form of the minimum may be determined when $\theta = \sigma$ and $u_i = v_i$. Then, since it is straightforward to show that

$$Q(\sigma,\sigma) = \frac{1}{\Omega} \int_{\Gamma_E} v_i \Pi^0_{ij} n_j ds, \tag{2.19}$$

we find the minimum is given by

$$-\frac{1}{\Omega} \int_{\Gamma_E} v_i \Pi^0_{ij} n_j ds = \text{minimum} . \tag{2.20}$$

For the slab geometry , $\Pi^0_{ij} n_j = -p_\pm \delta_{ij} n_j$ on the two bounding planes. Thus,

$$\frac{1}{\Omega} \int_{\Gamma_E} v_i \Pi^0_{ij} n_j ds = -\frac{\Delta p}{\Omega} \int_{\Gamma_+} v_i n_i ds \tag{2.21}$$

where the surface Γ_+ is the plane with pressure p_+ . We have used fluid incompressibility in simplifying (2.21); the volume of fluid flowing into the volume Ω_f across one plane must be matched by the volume of fluid flowing out of Ω_f across the other plane. Furthermore, the macroscopic flow U_z is defined by

$$U_z = \int_{\Gamma_+} v_i n_i ds/A \tag{2.22}$$

where, for an isotropic porous material with porosity (void volume fraction) ϕ , A is total area of a cross section of the porous medium and the volumes are given by $\Omega = A\Delta z$ and $\Omega_f = \phi\Omega$. Thus, in terms of macroscopic quantities, (2.20) becomes

$$\frac{\Delta P}{\Delta Z} U_z = \text{minimum}. \tag{2.23}$$

Now Darcy's law states that

$$U_z = - \frac{k}{\mu} \frac{\Delta P}{\Delta Z} \qquad (2.24)$$

where k is the permeability so (2.23) becomes

$$- \frac{k}{\mu} \left(\frac{\Delta P}{\Delta Z} \right)^2 = \text{minimum} \qquad (2.25)$$

for the variational problem (2.17). The zero upper bound on (2.24) implies a
zero lower bound on k.

The minimum for (2.18) follows from the preceding arguments by taking
$\tau = \sigma$. Then, we find

$$Q(\tau, \tau) > Q(\sigma, \sigma) = \frac{k}{\mu} \left(\frac{\Delta P}{\Delta Z} \right)^2 . \qquad (2.26)$$

Thus, the variational principle (2.18) may be rephrased as the statement that,
of all symmetric, zero trace tensors τ satisfying the admissibility conditions
(2.12) and (2.13), the unique choice $\tau = \sigma$ where σ satisfies (2.1) - (2.6)
gives the minimum energy dissipation in the fluid. This principle is the fluid-
dynamical version of Castigliano's principle for the equilibrium of an isotropic
elastic body while the principle (2.17) is the fluid version of the principle of
minimum potential energy in elasticity. The problem (2.18) is preferred for
viscous flow because a large class of trial stress distributions is easily
constructed as we will show in the next section.

To complete the formulation of the variational principle for Darcy's
constant, some method of introducing information about the random geometry of
the pore space must be provided. Following Prager [11], we introduce the
stochastic function

$$g(\vec{x}) = \begin{cases} 1 & \text{if } \vec{x} \in \Omega_f \\ \\ 0 & \text{if } \vec{x} \notin \Omega_f \end{cases} \qquad (2.27)$$

and the volume average

$$\langle \cdot \rangle = \frac{1}{\Omega} \int_\Omega \cdot\, d^3x \ . \tag{2.28}$$

Then, the quadratic form Q becomes

$$Q(\tau, \tau) = \frac{1}{2\mu} \langle g\, \tau_{ij}\tau_{ij} \rangle \ . \tag{2.29}$$

Comparing (2.26) and (2.29), we see that it would be advantageous to express the pressure gradient in terms of the trial stress. Such an expression may be obtained using the admissibility conditions (2.12) and (2.13). First, define the macroscopic pressure gradient

$$G_i = \delta_{i3}\, \frac{\Delta P}{\Delta Z} \ . \tag{2.30}$$

Then, consider

$$\langle g\tau_{ij,j} \rangle = \frac{1}{\Omega} \int_{\Omega_f} \tau_{ij,j}\, d^3x$$

$$= \frac{1}{\Omega} \int_{\Gamma_E} n_i\, p\, ds + \frac{1}{\Omega} \int_{\Gamma_I} n_i\, p\, ds \ . \tag{2.31}$$

Since the trial stress distribution satisfies (2.13) and since the trial pressure field $p = p^0$ on Γ_E , we have

$$\frac{1}{\Omega} \int_{\Gamma_E} n_i\, p\, ds = \delta_{i3}\, \frac{(P_+ - P_-)}{\Omega} \int_{\Gamma_E} ds = \phi G_i \ . \tag{2.32}$$

Furthermore, since $\tau_{ij} n_j = 0$ on Γ_E , the divergence theorem gives

$$\frac{1}{\Omega} \int_{\Omega_f} \tau_{ij,j}\, d^3x = \frac{1}{\Omega} \int_{\Gamma_I} \tau_{ij} n_j\, ds \ , \tag{2.33}$$

so the normalization condition may be expressed as

$$\frac{1}{\Omega} \int_{\Gamma_I} (\tau_{ij} - p\delta_{ij}) n_j\, ds = \phi G_i \ . \tag{2.34}$$

Eq. (2.34) is the correct normalization on a general trial stress τ . This normalization is the same one used by Weissberg and Prager [19] but it differs from the one used originally by Prager [11]. Ramifications of the switch from the erroneous normalization to the correct one have been discussed by Berryman and Milton [20]. Using (2.26), (2.29), (2.30), and (2.34), the variational principle may now be written as

$$k < \frac{1}{2} <g\tau_{ij}\tau_{ij}> /G_k G_k \; .$$

(2.35)

As a final check on the variational principle, we perform a first variation of (2.35). As formulated using (2.34), the variation of the denominator vanishes identically. The variation of the numerator is

$$\delta < g\tau_{ij}\tau_{ij}> = 2 < g\tau_{ij}\delta\tau_{ij}> \; .$$

(2.36)

If $\tau_{ij} = \mu(w_{i,j} + w_{j,i})$ for some vector w_i, then it follows easily that

$$< g\tau_{ij}\delta\tau_{ij} > = \frac{2\mu}{\Omega} \int_{\Gamma} w_i(\delta\tau_{ij} - \delta\rho\,\delta_{ij})\, n_j ds$$
$$+ \frac{2\mu}{\Omega} \int_{\Omega_f} w_{i,i}\, \delta\rho\, d^3x$$

(2.37)

The zero trace property of τ_{ij} guarantees that $w_{i,i} = 0$ in Ω_f so, if we define $\Pi_{ij} = -\rho\delta_{ij} + \tau_{ij}$, then

$$\delta < g\tau_{ij}\tau_{ij} > = \frac{4\mu}{\Omega} \int_{\Gamma_I} w_i\, \delta\Pi_{ij} n_j ds$$
$$+ \frac{4\mu}{\Omega} \int_{\Gamma_E} w_i\, \delta\Pi_{ij}^o\, n_j ds \; .$$

(2.38)

The second surface integral in (2.38) vanishes because Π_{ij}^o is specified on the external boundary. The variation (2.34) therefore vanishes only if $w_i = 0$ on Γ_I since the variations $\delta\Gamma_{ij}$ may be arbitrary there. Thus, the ratio in (2.35) is stationary if the admissible trial stress τ satisfying (2.12) and (2.13) also satisfies (2.10) and (2.11). A second variation of (2.35) produces a positive result; hence, the stationary point is in fact a minimum.

This calculation concludes the formulation of the variational principle. The next section discusses methods of choosing useful trial stress distributions.

3. Trial Stress Distributions

Trial stress distributions τ for the variational bound (2.35) on Darcy's constant must satisfy the following admissibility conditions:

$$\tau_{ij} = \tau_{ji} \, , \tag{3.1}$$

$$\tau_{kk} = 0 \, , \qquad \text{in } \Omega_f \, , \tag{3.2}$$

$$\tau_{ij,j} = p_{,i} \tag{3.3}$$

$$\tau_{ij} n_j = (\pi_{ij}^0 + p^0 \delta_{ij}) n_j \quad \text{on } \Gamma_E \, , \tag{3.4}$$

and

$$\frac{1}{\Omega} \int_{\Gamma_I} (\tau_{ij} - p\delta_{ij}) n_j \, ds = \phi \frac{\Delta P}{\Delta Z} \delta_{i3} \, . \tag{3.5}$$

We may replace the admissibility condition (3.3) by the equivalent condition

$$\varepsilon_{mik} \tau_{ij,jk} = 0 \, . \tag{3.6}$$

In (3.6), ε_{mik} is the Levi-Civita symbol (defined to be zero if any two of the indices are equal and either +1 or -1 for even or odd permutations of 123) so (3.3) has been replaced by the statement that the curl of a gradient vanishes identically. The principal effect of (3.5) is to show that τ and p must be correlated with the stochastic function q and the unit outward normal vector n_i . However, the required correlation of q and τ is not a very stringent condition; many examples of suitable choices for τ could be listed including various functions and functionals of q.

The most serious difficulty with the application of the variational principle is incomplete knowledge of q. In computer experiments on flow through random aggregates, it is possible to have as much information as desired about the stochastic function. In any other circumstance of practical interest, we should assume from the outset that only limited knowledge of the statistical properties of the porous medium will be available and design our trial functions to use the information at hand. Prager [11] provided one solution to this problem by introducing the trial stress distribution

$$\tau_{ij}(\vec{x}) = \int T_{ij}(\vec{r}) \, h(\vec{x} + \vec{r}) \, d^3r \ , \tag{3.7}$$

where T_{ij} is a deterministic tensor and h is some stochastic scalar while the integral is over all space. The deterministic part of this functional $T_{ij}(\vec{r})$ may then be required to satisfy

$$T_{ij} = T_{ji} \ , \tag{3.8}$$

$$T_{kk} = 0 \ , \tag{3.9}$$

and

$$T_{ij,j} = \psi_{,i} \tag{3.10}$$

for some deterministic scalar function ψ or equivalently

$$\varepsilon_{mik} \, T_{ij,jk} = 0 \tag{3.11}$$

in order to guarantee satisfaction of (3.1), (3.2), and (3.6). Prager made the particularly simple choice

$$h(\vec{x}) = g(\vec{x}) \tag{3.12}$$

for the stochastic scalar; however, we will show in the following sections that other choices of h are both possible and preferable in some cases. Along with the choice (3.12), we make the assumption that $g(x)$ may be extended outside the volume Ω so that the volume integral in (3.7) may be evaluated. A periodic extension of g is conceptually simple but undesirable because of the long range order introduced this way. A superior choice of extension is to assume that other samples with matching pore structure are drawn from the same ensemble as g and arranged to fill all space. This extension causes no conceptual or practical difficulties. We must also assume the scale of the porous microstructure is much smaller than the thickness of the sample so the satisfaction of the surface boundary condition (3.4) is assured.

Using linearity with respect to the applied pressure gradient and the symmetric nature of T from (3.8) while assuming that the medium is isotropic, Prager obtained the general form of the deterministic function. In general, this result becomes

$$T_{ij}(\vec{r}) = \{\alpha(r)\,[r_j G_i + r_i G_j]$$

$$+ \beta(r) G_k r_k r_i r_j + \gamma(r) G_k r_k \delta_{ij}\} \;.$$
(3.13)

The scalar functions α , β , and γ depend only on the magnitude $r = |\vec{r}|$. The tensor (3.13) will satisfy (3.9) if

$$\gamma(r) = -\frac{1}{3}[2\alpha(r) + r^2\beta(r)] \;.$$
(3.14)

T will also satisfy (3.11) if, for $i \neq k$,

$$0 = T_{ij,jk} - T_{kj,ji}$$
(3.15)

$$= (\alpha'' + \frac{4\alpha'}{r} - 5\beta - r\beta')(G_i r_k - G_k r_i).$$

Thus, we find that

$$\frac{d}{dr}[r^5\beta(r)] = \frac{d}{dr}[r^4\frac{d\alpha(r)}{dr}]$$
(3.16)

or equivalently that

$$r^5\beta(r) = r^4\frac{d\alpha(r)}{dr} + \text{const.}$$
(3.17)

Except for the constant appearing in (3.17), β and γ are completely determined by (3.17) and (3.14) once a functional form for α has been chosen. Several choices for the function α will be discussed in Section 5.

The constant in (3.17) is fixed by the normalization condition (3.5). To see this, substitute (3.13) into (3.10) and note [20] that a solution for ψ is

$$\psi(\vec{r}) = G_k r_k [r\frac{d\alpha(r)}{dr} + \frac{10}{3}\alpha(r) - \frac{1}{3}r^2\beta(r)] \;.$$
(3.18)

Furthermore, the trial pressure field ρ is given by

$$\rho(\vec{x}) = \int \psi(\vec{r}) \ h(\vec{x} + \vec{r}) \ d^3r \ . \qquad (3.19)$$

After some simplification, we find [20] that (3.5) becomes

$$\int d^3r [T_{ij}(\vec{r}) - \psi(\vec{r})\delta_{ij}] \frac{r_j}{r} \frac{dS_2(r)}{dr} \qquad (3.20)$$

$$= \phi G_i$$

where $S_2(r)$ is the two-point (auto-) correlation function of $g(x)$ satisfying $S_2(0) = \phi$ and $S_2(\infty) = \phi^2$. Then Eq. (3.20) reduces to the statement that

$$\frac{1}{3} \int d^3r \ r^2 [r\beta(r) - \frac{d\alpha(r)}{dr}] \frac{dS_2(r)}{dr} \qquad (3.21)$$

$$= - \frac{4\pi}{3} \phi(1-\phi) \ \text{const} = \phi \qquad (3.22)$$

using (3.17). So the constant in (3.17) is just

$$\text{const} = - \frac{3}{4\pi(1-\phi)} \ .$$

4. Phase Interchange Relation

In this section, we will use the flexibility in choice of stochastic function in (3.7) to derive a phase interchange relation for permeability. By a phase interchange relation, we mean any equality or inequality which applies to a sum or product of the effective constants when the geometry of the material is held fixed but the roles of the constitutents are interchanged. Keller [23] derived such a phase interchange equality for the products of conductivities of two-phase composites in two dimensions. Keller's results were generalized to an inequality for products of conductivities of two-phase composites in three dimensions by Schulgasser [24]. We will derive an inequality for a linear combination of the permeabilities for a porous medium and its complement.

The complementary medium has the same geometrical structure as the original medium but the locations of solid and void are interchanged. As a mathematical consequence of this interchange, the stochastic function $\overline{g}$ for the complement is

$$\overline{g}(\vec{x}) = 1 - g(\vec{x}) \tag{4.1}$$

if the stochastic function g for the original medium is given by (2.27). Using the bar to distinguish properties of the complementary medium, we find easily that the porosity of the complement is

$$\overline{\phi} = 1 - \phi \tag{4.2}$$

and we define the complementary permeability to be $\overline{k}$.

Now it will prove to be illuminating to consider trial stress distributions of the form (3.7) for the original porous medium with either of two choices for h

$$h_1(\vec{x}) = g(\vec{x}) \tag{4.3}$$

or

$$h_2(\vec{x}) = \overline{g}(\vec{x}) \tag{4.4}$$

and for T_{ij} we choose the (so-called) constant trial function given by (3.17) with $\alpha = 0$. Making these substitutions into (2.31), we find that the results depend on certain spatial correlation functions of the form

$$C_2^{(m)}(\vec{r}) = \langle g(\vec{x})h_m(\vec{x}+\vec{r})\rangle \tag{4.5}$$

and

$$C_3^{(m)}(\vec{r},\vec{s}) = \langle g(\vec{x})h_m(\vec{x}+\vec{r})h_m(\vec{x}+\vec{s})\rangle \ . \tag{4.6}$$

The two variational bounds associated with the (4.3) and (4.4) are then

$$k < \iint T_{ij}(\vec{r})T_{ij}(\vec{s})C_3^{(m)}(\vec{r},\vec{s})d^3r\,d^3s/2G_k G_k \tag{4.7}$$

for m=1 and 2.

Two important identities follow for (4.3) - (4.6):

$$c^{(2)}_{2,j} (\vec{r}) = - c^{(1)}_{2,j} (\vec{r}) \tag{4.8}$$

and

$$c^{(2)}_3 (\vec{r},\vec{s}) = \phi - c^{(1)}_2 (\vec{r}) - c^{(1)}_2 (\vec{s}) + c^{(1)}_3 (\vec{r},\vec{s}) \ . \tag{4.9}$$

If we substitute (4.9) into the numerator of (4.7) for $m=2$, we find that the first three terms on the right hand side of (4.9) do not contribute to the double integral if T_{ij} is of the form (3.13) since it is easily shown that

$$\int T_{ij}(\vec{r})d^3r = 0 \ . \tag{4.10}$$

The physical significance of (4.10) is that the mean stress deviations must vanish $\langle \tau_{ij} \rangle = 0$ since the system as a whole is not being sheared. Thus, the numerators of both bounds are equal and the bounds for either choice (4.3) or (4.4) are the same. This result shows that the bounds depend most strongly on the arrangement of the internal surface which is identical for the sample and its complement.

Now consider the related spatial correlation functions

$$\overline{C}^{(m)}_2 (\vec{r}) = \langle \overline{g}(\vec{x})h_m(\vec{x}+\vec{r}) \rangle \tag{4.11}$$

and

$$\overline{C}^{(m)}_3 (\vec{r},\vec{s}) = \langle \overline{g}(\vec{x})h_m(\vec{x}+\vec{r})h_m(\vec{x}+\vec{s}) \rangle . \tag{4.12}$$

Then it is straightforward to show that

$$\overline{C}^{(m)}_{2,j} (\vec{r}) = - c^{(m)}_{2,j} (\vec{r}) \tag{4.13}$$

and

$$\overline{C}^{(m)}_3 (\vec{r},\vec{s}) + c^{(m)}_3 (\vec{r},\vec{s}) = \langle h_m(\vec{x}+\vec{r})h_m(\vec{x}+\vec{s}) \rangle$$

$$\equiv S^{(m)}_2 (|\vec{r}-\vec{s}|) \ . \tag{4.14}$$

68

Using (4.13), the phase-interchanged version of (4.7) becomes

$$\overline{k} \leq \left(\frac{\overline{\phi}}{\phi} \right)^2 \iint T_{ij}(\vec{r})T_{ij}(\vec{s})\overline{C}_3^{(m)}(\vec{r},\vec{s})d^3r\ d^3s/2G_k G_k \ , \qquad (4.15)$$

where we have accounted for the difference in normalization constant by the factor preceding the integral. Multiplying (4.15) by $(\phi/\overline{\phi})^2$, adding the result to (4.7) and using (4.14), we have

$$k + \left(\frac{\phi}{\overline{\phi}} \right)^2 \overline{k} \leq \phi^2 \iint T_{ij}(\vec{r})\ T_{ij}(\vec{s})\ S_2^{(m)}\ (|\vec{r}-\vec{s}|)d^3r\ d^3s/2G_k G_k \ . \qquad (4.16)$$

The inequality (4.16) is a phase interchange relation for this linear combination of the permeability k and its complementary value $\overline{k}$.

To begin analyzing (4.16), first notice that this upper bound depends only on two-point spatial correlation functions. Next notice that, since both contributions to the left hand side of (4.16) are strictly positive, we may eliminate either term and still have a valid upper bound on the remaining term. Thus, in (4.7) $C_3^{(m)}$ may be replaced by $S_2^{(m)}$ while maintaining the validity of the inequality. Such a replacement has some advantages for practical applications because two-point spatial correlation functions are more easily measured or computed than three-point correlations.

The two-point upper bound in (4.16) is not a new result. Prager [11] showed that, since g ≤ 1 , the factor of g in the numerator of (2.35) can be replaced by unity and still maintain a valid bound on Darcy's constant. The result is

$$k \leq <\tau_{ij}\tau_{ij}>/2G_k G_k \ , \qquad (4.17)$$

which has the same right hand side as (4.16). The new result (4.16) therefore shows that the bound (4.17) <u>must</u> be a poor estimate of k if $\overline{k}$ is positive. If ϕ is large, then it will often be the case that there are no connecting paths through the particle-phase; then the permeability $\overline{k}$ for the interchanged problem will vanish identically. Thus, it is possible for (4.17) to provide a good estimate of k when $\phi \to 1$, as has been observed in numerical calculations [13-15,20]. Similarly, if ϕ is small, then it is possible that no connecting paths through the void-phase exist; then the permeability k will vanish identically. But (4.16) shows that the right hand side of (4.17) is bounded away from zero even when $k = 0$ because $\overline{k}$ will be positive (and large) in this limit. This argument shows why the two-point bounds (4.17) (which are evaluated numerically in the next section) provide such bad estimates of k at low porosities.

5. Two-Point Bounds

Attention will now be restricted to two-point bounds to provide some elaboration of the results on the phase interchange relation (4.16). It has been shown elsewhere [20] that the best possible two-point bound obtainable using the stress distribution (3.7) together with (3.12) has the form

$$k < \frac{2}{3} \int_0^\infty dr \; r[S_2(r) - \phi^2]/(1-\phi)^2 \, , \tag{5.1}$$

with $S_2 = S_2^{(1)}$ defined in (4.14). The relation (5.1) was found by noting that, if the deterministic function T_{ij} is allowed to vary over the entire class of admissible tensors, the smallest value of $1/2 < \tau_{ij}\tau_{ij} >$ is the right hand side of (5.1).

To permit the evaluation of (5.1), the discussion must also be limited to a particular kind of random geometry for which the various two-point spatial correlation functions are known. Until such data become available for real materials [16], we are essentially limited to two types of model materials: (1) random packings of impenetrable spheres and (2) random assemblages of penetrable spheres. Two-point bounds for impenetrable-sphere packs have been discussed in [14]. For the present discussion, we restrict our analysis to the penetrable sphere model.

The penetrable sphere model assumes that particle centers are distributed randomly in the volume Ω and that each center is surrounded by a sphere of particle material. In the simplest version of this model, all these spheres have the same radius R. If the density of particle centers is great enough and the sphere radius is large enough the spheres will overlap. This model has the distinct advantage that analytical results are known for all the spatial correlation functions of interest [12,15,25-27].

Since the particle centers are uncorrelated, it is not difficult to show [25-27] that the general result for an n-point void correlation function is

$$S_n(\vec{x}_1,\ldots,\vec{x}_{n-1}) \equiv \langle g(\vec{x})g(\vec{x}+\vec{x}_1)\ldots g(\vec{x}+\vec{x}_{n-1})\rangle = \exp(-\rho V_n) \qquad (5.2)$$

where ρ is the number density of spheres and V_n is the union volume of n spheres with the fixed radius R and centers at the vertices $\vec{x}_1,\ldots,\vec{x}_{n-1}$. The union volume for one sphere is just

$$V_1 = \frac{4\pi}{3} R^3 , \qquad (5.3)$$

so the porosity is given by

$$\phi = S_1 = \exp(-\rho\frac{4\pi}{3} R^3) . \qquad (5.4)$$

For two spheres, the union volume is found to be

$$V_2(\zeta R)/R^3 \;=\; \begin{cases} \dfrac{4\pi}{3}\left(1 + \dfrac{3}{4}\zeta - \dfrac{\zeta^3}{16}\right) & \text{for } \zeta \leqslant 2 \\[2ex] \dfrac{8\pi}{3} & \text{for } \zeta \geqslant 2 \end{cases} \qquad (5.5)$$

and S_2 follows from (5.2).

The results of computations using the penetrable sphere model to provide the data needed for the two-point correlation functions are summarized in Table I and Figure 1. For comparison, the analytical result of Weissberg and Prager [19] that

$$k \leqslant -\frac{2}{9} R^2 \frac{\phi}{\ln\phi} \qquad (5.6)$$

is also shown in Figure 1. The inequality in (5.6) was derived for the penetrable sphere model using the same variational principle discussed in Section 2, but a different trial function which takes advantage of the symmetries inherent in assemblages of penetrable spheres.

Also shown is the low density expansion for the permeability of an assemblage of hard spheres [6-8]

$$k = \frac{2}{9} \frac{R^2}{\eta} \left[1 + \frac{3}{\sqrt{2}} \eta^{1/2} + \frac{135}{64} \eta \log \eta + 16.5\eta + ... \right]^{-1} \qquad (5.7)$$

where the solid volume fraction is $\eta = 1 - \phi$.

Table I and Figure 1 show that the result of Weissberg and Prager is somewhat superior to the two-point bound of Prager for the penetrable sphere model. However, two distinctions should be stressed: (1) The variational bound (5.1) is valid for any isotropic porous solid whereas (5.6) is valid only for the penetrable sphere model. (2) Because the derivation of (5.6) differs significantly from that of (5.1), it cannot be said that (5.6) is a "two-point bound". Indeed the derivation of (5.6) avoids the introduction of the spatial correlation functions altogether. Unfortunately, the trick which leads to (5.6) appears to work only for the penetrable sphere model, so (5.1) appears to be the best <u>general</u> bound on permeability using the limited information in the two-point correlation functions.

6. Discussion

As a final example of the versatility of the variational principle (2.35), consider two porous media with stochastic functions f and g whose porosities are respectively

$$\langle f \rangle = \phi_f \quad \text{and} \quad \langle g \rangle = \phi_g \ . \qquad (6.1)$$

Then, if we define the right hand side of (2.35) to be $k_g(\tau)$, we have

$$k \le k_g(\tau) = \langle g\tau_{ij}\tau_{ij}\rangle / 2G_k G_k \ . \qquad (6.2)$$

Now suppose that the stochastic functions of these two materials are related by the inequality

$$f(\vec{x}) \ge g(\vec{x}) \quad \text{for all} \ \vec{x} \ , \qquad (6.3)$$

so the material characterized by f is more porous than the one for g; in particular, the material with stochastic function g is obtained by adding more solid material to the other one without rearranging the original material. Then, it is clearly true that

$$\langle g\tau_{ij}\,\tau_{ij}\rangle \; \leq \; \langle f\tau_{ij}\tau_{ij}\rangle \tag{6.4}$$

for any given trial stress distribution τ . Substituting (6.4) into (6.2) and defining $k_f(\tau)$ as the right hand side of (2.35) with g replaced by f , we find

$$k \; \leq \; k_g(\tau) \; \leq \; k_f(\tau) \tag{6.5}$$

where k is the actual permeability for the material characterized by g .

The inequalities in (6.5) are true for any fixed τ . In particular, if $\tau = \sigma_f$ where σ_f minimizes k_f , then

$$k \; \leq \; k_g\,(\sigma_f) \; \leq \; k_f(\sigma_f) \; . \tag{6.6}$$

Similarly, if σ_g minimizes k_g , then

$$k \; = \; k_g\,(\sigma_g) \; \leq \; k_f(\sigma_g) \; . \tag{6.7}$$

Thus, it follows from (6.6) and (6.7) that

$$k \; = \; k_g(\sigma_g) \; \leq \; k_g(\sigma_f) \; \leq \; k_f(\sigma_f) \; . \tag{6.8}$$

This argument shows that, if one porous material differs from a more porous one only by the addition of solid material (with no rearrangement of the other material), then the permeability of the more porous material is always greater than or equal to that of the less porous one.

The inequalities in (6.5) have other practical consequences. If we use the same trial function τ for two problems with f and g related by (6.3), then the bound obtained for the more porous material will be a valid bound for the less porous one. This relationship is observed to be satisfied by the examples in Table I for the penetrable sphere model. Specific representations of the penetrable sphere model can satisfy (6.3) very easily by choosing a particular set of sphere centers and letting the radii of the overlapping spheres satisfy $R_f < R_g$ for the two cases. Thus, (6.5) is a useful check on our numerical integration method and also guarantees that the curves in Figure 1 will decrease monotonically as observed.

In conclusion, the main result of the present work is the observation that the stochastic function h appearing in the trial stress distribution (3.8) can take a variety of forms. We have used the freedom in choosing both the deterministic function T_{ij} and the stochastic function h in (3.8) to obtain several new results. A phase interchange relation for a linear combination of the permeability and its complementary value was obtained in (4.16). A proof of the monotone properties of the bounds was given in (6.8). Numerical comparisons of bounds on permeability were made for the penetrable sphere model once the limited geometrical information contained in the two-point spatial correlation functions is available. Future developments in the theory will require knowledge of three-point and possibly higher order spatial correlation functions.

Acknowledgments

I thank Graeme W. Milton for enlightening correspondence which helped to elucidate the correct form of the normalization condition in the variational principle. This work was performed under the auspices of the U.S. Department of Energy by the Lawrence Livermore National Laboratory under contract number W-7405-ENG-48 and supported specifically by the Nuclear Test Containment Program.

References

1. H. Darcy, "Les fontaines publique de la ville de Dijon," Paris, 1856.

2. M. Poreh and C. Elata, "An analytical derivation of Darcy's law," Israel J. Tech. $\underline{4}$, 214-217 (1966).

3. S.P. Neuman, "Theoretical derivation of Darcy's law," Acta Mech. $\underline{25}$, 153-170 (1977).

4. J.B. Keller, "Darcy's law for flow in porous media and the two-space method," in _Nonlinear Partial Differential Equations in Engineering and Applied Science_, ed. by R.L. Sternberg, A.J. Kalinowski, and J. S. Papadakis (Marcel Dekker, New York, 1980), pp. 429-443.

5. H.C. Brinkman, "A calculation of the viscous force exerted by a flowing fluid on a dense swarm of particles," Appl. Sci. Res. $\underline{A1}$, 27-34 (1947).

6. S. Childress, "Viscous flow past a random array of spheres," J. Chem. Phys. $\underline{56}$, 2527-2539 (1972).

7. I.D. Howells, "Drag due to the motion of a Newtonian fluid through a sparse random array of small fixed rigid objects," J. Fluid Mech. $\underline{64}$, 449-475 (1974).

8. E.J. Hinch, "An averaged-equation approach to particle interactions in a fluid suspension," J. Fluid Mech. $\underline{83}$, 695-720 (1977).

9. A.A. Zick and G.M. Homsy, "Stokes flow through a periodic array of spheres," J. Fluid Mech. $\underline{115}$, 13-26 (1982).

10. A.S. Sangani and A. Acrivos, "Slow flow through a periodic array of spheres," Int. J. Multiphase Flow $\underline{8}$, 343-360 (1982).

11. S. Prager, "Viscous flow through porous media," Phys. Fluids $\underline{4}$, 1477-1482 (1961).

12. M. Doi, "A new variational approach to the diffusion and the flow problem in porous media," J. Phys. Soc. Japan $\underline{40}$, 567-572 (1976).

13. S. Torquato, "Microscopic approach to transport in two-phase random media," Ph.D. thesis (State University of New York at Stony Brook, 1980).

14. J.G. Berryman, "Computing variational bounds for flow through random aggregates of spheres," J. Comput. Phys. $\underline{52}$, 142-162 (1983).

15. J.G. Berryman, "Bounds on fluid permeability for viscous flow through porous media," J. Chem. Phys. $\underline{82}$, 1459-1467 (1985).

16. J.G. Berryman, "Measurement of spatial correlation functions using image processing techniques," J. Appl. Phys. $\underline{57}$, 2374-2384 (1985).

17. R. Courant and D. Hilbert, _Methods of Mathematical Physics_, Vol. I, (Interscience, New York, 1953), pp. 252-257 and 268-272.

18. M.J. Beran, _Statistical Continuum Theories_ (Interscience, New York, 1968),
Chapter 6.

19. H.L. Weissberg and S. Prager, "Viscous flow through porous media. III. Upper
bounds on the permeability for a simple random geometry," Phys. Fluids
$\underline{13}$, 2958-2965 (1970).

20. J.G. Berryman and G.W. Milton, "Normalization constraint for variational
bounds on fluid permeability", J. Chem. Phys., July, 1985.

21. L.D. Landau and E.M. Lifshitz, _Fluid Mechanics_ (Pergamon Press, London,
1959), p. 54.

22. H. Lamb, _Hydrodynamics_ (Dover, New York, 1945), pp. 617-619.

23. J.B. Keller, "A theorem on the conductivity of a composite medium," J.
Mathematical Phys. $\underline{5}$, 548-549 (1964).

24. K. Schulgasser, "On a phase interchange relationship for composite
materials," J. Mathematical Phys. $\underline{17}$, 378-381 (1976).

25. H.L. Weissberg, "Effective diffusion coefficient in porous media," J. Appl.
Phys. $\underline{34}$, 2636-2639 (1963).

26. W. Strieder and R. Aris, _Variational Methods Applied to Problems of
Diffusion and Reaction_ (Springer-Verlag, New York, 1973), pp. 4-8.

27. S. Torquato and G. Stell, "Microstructure of two-phase random media. III.
The n-point matrix probability functions for fully penetrable spheres,"
J. Chem. Phys. $\underline{79}$, 1505-1510 (1983).

η	k_{WP}/R^2	k_p/R^2
0.1	1.898E+00	2.329E+00
0.2	7.967E-01	1.001E+00
0.3	4.361E-01	5.611E-01
0.4	2.610E-01	3.443E-01
0.5	1.603E-01	2.170E-01
0.6	9.701E-02	1.348E-01
0.7	5.537E-02	7.885E-02
0.8	2.761E-02	4.009E-02
0.9	9.651E-03	1.395E-02

Table I. Values of the upper bounds on permeability due to Weissberg and Prager (k_{WP}) and due to Prager (k_p). The corresponding formulas in the text are respectively (5.6) and (5.1). The solid volume fraction $\eta = 1-\phi$.

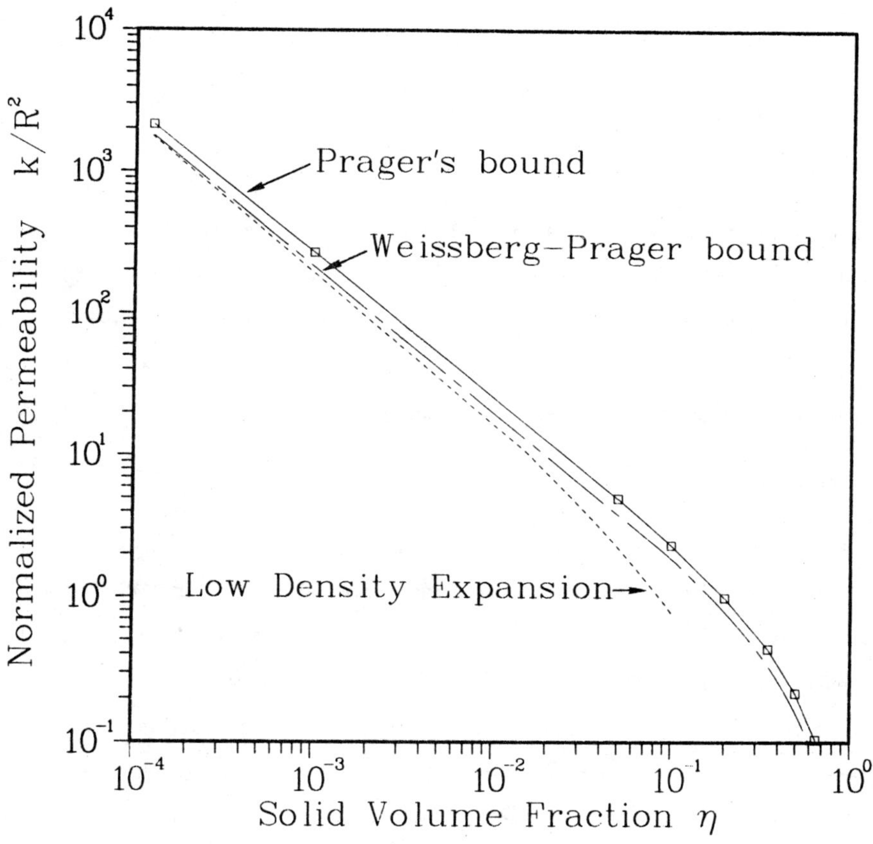

Figure 1. Comparison of Prager's two-point bound (boxes and solid line) to Weissberg and Prager's analytical bound (long-dash/short-dash line) for penetrable spheres and to the low density expansion for hard spheres (dashed line) as the porosity ϕ or solid volume fraction $\eta = 1-\phi$ varies. The corresponding equations in the text are respectively (5.1), (5.6), and (5.7). Some numerical values are also listed in Table I.

MICROMODELING OF VOID GROWTH AND COLLAPSE

M.M. Carroll

Department of Mechanical Engineering
University of California, Berkeley 94720

Abstract

Use of the hollow sphere model to describe the response of porous materials and powders under tensile or compressive stress is described. Topics include pore pressure effects, powder compaction equations, rate effects, load maxima, and deviatoric effects. A recent hybrid empirical-microstructural model for creep compaction of heated metal powders (HIP) is described in some detail.

1. Introduction

A porous material may be modeled as an effectively homogeneous continuum of the usual kind or as a generalized continuum, with additional kinematical variables to describe the pore space distribution. The latter approach, with a single scalar variable to describe the volume porosity, was used by Herrmann [1] for porous metals and by Goodman and Cowin [2] for granular materials. A variant of the generalized continuum approach is the use of the theory of mixtures (interacting continua), and this may be especially useful in describing the behavior of fluid-filled porous solids.

It is debatable whether or not a single additional scalar variable provides an adequate description of the pore space distribution. One might generate tensor measures by defining line and area porosities as well as the volume porosity. However, if the pore space distribution is statistically homogeneous and isotropic, then all three measures will take the same value (see Martin [3] for a discussion of area and volume porosities). Another approach is to introduce various moments of the pore space distribution (cf. Drew [4]). There has been a considerable discussion of tensor measures of fabric in granular materials (see, for example, Jenkins and Satake [5]). A tensor measure of the change in pore space arises from a discussion of volume average strains in a porous material, as discussed in the following section.

Useful surveys of work on granular, porous or heterogeneous media are contained in the Seminar Proceedings edited by Jenkins and Satake [5], the Symposium Proceedings edited by Cowin and Carroll [6] and in special journal issues on heterogeneous materials [7,8]. In particular, these surveys present treatments of porous materials either as assemblages of grains in contact or as distributions of voids in a solid matrix. Sometimes both of these approaches are used, as in descriptions of metal powder compaction which model the initial response in terms of grain contact and the later stage in terms of stress concentrations around voids.

In the present paper, we describe an approach to the development of constitutive equations for porous materials which is in part micromechanical and in part empirical. Thus, it makes use of solutions of boundary value problems for an idealized micromodel - the spherical pore model - in detail. We present a constitutive theory for hot <u>isostatic</u> (homothermal, volumetric creep) compaction of metal powders. We also treat thermal effects in shock compaction.

2. General Results

Some general results on the behavior of porous materials may be obtained by applying well-known theorems on volume averages of stress and strain. For example, the volume average $\overline{\sigma}_{ij}$ of the Cauchy stress in a body which is in equilibrium under no body force and occupies a region R with boundary B is given by

$$\overline{\sigma}_{ij} = \frac{1}{V} \int_B t_i x_j \, da \quad , \tag{1}$$

where V is the volume of R , t_i is the surface traction on B and x_j are rectangular Cartesian coordinates. This result is an immediate consequence of the equation of equilibrium and it is noteworthy that the applied load and the current geometry determine the average stress independently of the material response. Similarly, the volume average $\overline{e}_{ij}$ of the infinitesimal strain in the body is given by

$$\overline{e}_{ij} = \frac{1}{2V} \int_B (u_i n_j + u_j n_i)\,da \quad , \tag{2}$$

where u_i are displacement components and n_i are components of the unit outward normal on B. Equation (2) shows that the average strain is determined by the boundary displacements and, in particular, this allows us to define the "strain" of a void.

Expressions similar to equations (1) and (2) may be obtained for the averages of the Piola stress, the deformation gradients, the rate of deformation tensor, etc.

Equations (1) and (2) may be used to develop a theory for the static response of a fluid-filled porous elastic solid (Carroll and Katsube [9]). The governing equations may be summarized as follows:

$$e_{ij} = e^S_{ij} + e^*_{ij} \quad ; \quad \sigma_{ij} = \sigma^S_{ij} - \frac{\phi}{1-\phi} \sigma^*_{ij} \quad , \tag{3}$$

and

$$e^S_{ij} = C^S_{ijk\ell}\, \sigma^S_{k\ell} \quad ; \quad e^*_{ij} = C^*_{ijk\ell}\, \sigma^*_{k\ell} \quad . \tag{4}$$

The porosity ϕ is the ratio of pore volume to total volume, ie.,

$$\phi = V^p/V \quad , \tag{5}$$

σ_{ij} and e_{ij} denote the stress and strain in the porous material and σ^S_{ij} and e^S_{ij} the corresponding quantities for the solid matrix. The differential strain e^*_{ij} is defined by

$$e^*_{ij} = \phi(e^p_{ij} - e^S_{ij}) \quad , \tag{6}$$

where e^p_{ij} denotes the strain of the pore space. The Terzaghi effective stress, σ^*_{ij} is given by

$$\sigma_{ij}^* = \sigma_{ij} + p\delta_{ij} \quad , \tag{7}$$

where p is the pore fluid pressure. It is noteworthy that e_{ij}^* provides a tensor measure of the change in pore geometry and that this change is effected by the Terzaghi effective stress, as shown by the second of equations (4). The tensor $C_{ijk\ell}^S$ is the elastic compliance tensor for the solid material, while $C_{ijk\ell}^*$ governs the change in relative strain. The overall response is governed by the constitutive equation

$$e_{ij} = C_{ijk\ell}\hat{\sigma}_{k\ell} \quad , \tag{8}$$

with effective compliance tensor $C_{ijk\ell}$ given by

$$C_{ijk\ell} = \frac{1}{1-\phi} C_{ijk\ell}^S + C_{ijk\ell}^* \tag{9}$$

and "effective stress" tensor $\hat{\sigma}_{ij}$ given by

$$\hat{\sigma}_{ij} = \sigma_{ij} + P(\delta_{ij} - M_{ijk\ell}C_{k\ell mm}^S) \quad . \tag{10}$$

The effective compliance tensor $C_{ijk\ell}$ is more readily measurable experimentally than $C_{ijk\ell}^*$, which can be determined from equation (9). The tensor $M_{ijk\ell}$ is the effective modulus tensor and equation (10) provides an "effective stress law" for anisotropic elastic deformation.

Remark:

Equation (5) defines one of several measures of porosity. The measure

$$n = V^p/V_o \quad , \tag{11}$$

where the subscript o denotes initial value, is commonly used in soil mechanics but it is not as convenient as the measure ϕ for large volume strains. The measure

$$\alpha = V/V^S = 1/(1-\phi) \tag{12}$$

is commonly used by shock wave physicists and the "relative density"

$$D = V^S/V = 1-\phi \tag{13}$$

is used by powder metallurgists.

3. The Hollow Sphere Model

The hollow sphere model was apparently first introduced by Torre [10] in 1948 and it has been widely used since then as an appropriate micromodel for porous material response. Torre's application was to obtain a compaction equation for metal powders. A similar model was used subsequently by Mackenzie and Shuttleworth [11], to model linearly viscous response, and by Mackenzie [12] to obtain effective moduli for an isotropic poroelastic material in the linear response range.

A major simplification follows from the fact that the relation between pressure and porosity does not depend on the material compressibility. This allows idealization of the material as incompressible, which reduces the spherically symmetric problem to a one-degree-of-freedom system, so that solutions may be obtained for materials with fairly complicated response and which include rate effects. Because of the simplicity of the model, its considerable success in fitting a great variety of data is rather surprising. This success may be explained, to some extent, by appealing to Hashin's composite sphere model [13].

(1) <u>Basic equations</u>

Consider a volume preserving, spherically symmetric deformation described by

$$r^3 - a^3 = r_0^3 - a_0^3 \quad ; \quad \theta = \theta_0 \quad ; \quad \phi = \phi_0 , \tag{14}$$

where (r_0, θ_0, ϕ_0) and (r, θ, ϕ) denote spherical polar coordinates of a particle before and after deformation and a_0 and a denote initial and present values of the inner radius of a hollow sphere. The local deformation consists of a radial stretch λ $(= r_0^2/r^2)$, with equal lateral stretches $\lambda^{-1/2}$; the radial stretch is extensional if the sphere is compacting, and vice versa. The local state of Cauchy (true) stress is a uniaxial radial stress together with a hydrostatic stress, and the latter has no effect because of incompressibility. It follows that the relevant material response property for spherical compaction is the axial stress-strain relation for uniaxial tension and that for spherical inflation is the axial stress-strain relation for uniaxial compression.

For the quasi-static compaction problem, the governing equations are the radial equation of equilibrium

$$\frac{d\sigma_{rr}}{dr} + \frac{2}{r}(\sigma_{rr} - \sigma_{\theta\theta}) = 0$$

and the conditions at the inner and outer boundaries

$$\sigma_{rr} = 0 \text{ at } r = a \text{ ; } \sigma_{rr} = -P \text{ at } r = b \tag{15}$$

supplemented by the appropriate response law

$$\sigma = \hat{\sigma}(\lambda) = \tilde{\sigma}(\varepsilon) \quad (\varepsilon = \ln\lambda) \tag{16}$$

under uniaxial tension σ . Equations (14)-(16) lead to

$$P = 2 \int_a^b \hat{\sigma}(\lambda) \frac{dr}{r} \text{ ; } \lambda = (r_0/r)^2 \text{ ,} \tag{17}$$

with r_0 given by the first equation (14). The initial and present values of porosity ϕ of the sphere are

$$\phi = (a/b)^3 \text{ ; } \phi_0 = (a_0/b_0)^3 \tag{18}$$

and corresponding values of α and D may be read off from equations (12) and (13).

(2) Powder compaction

Konopicky [14] and Shapiro and Kolthoff [15] introduced an empirical equation of the form

$$P = - A + B \ln \frac{1}{1-D} \tag{19}$$

to describe the relationship between pressure and relative density during static compaction of metal powders. This provides an excellent fit for many different materials and it has been widely used by powder metallurgists. Torre [10] solved the problem of compaction for a hollow sphere of rigid-perfectly plastic material.

84

Setting $\delta(\lambda) = Y$ (constant) in equation (17) gives

$$P = \frac{2}{3} Y \ln \frac{1}{1-D} \quad , \tag{20}$$

which is somewhat similar to the empirical equation (19). Carroll and Kim [16] attempted to find simple compaction equations for strain hardening materials by choosing forms of the strain hardening law in simple tension for which equation (17) can be integrated in terms of elementary functions. In particular, they studied the saturation hardening of Voce [17] and Palm [18], which has the form

$$\sigma = Y_\infty - (Y_\infty - Y_0)e^{-\varepsilon/\varepsilon_c} \quad , \tag{21}$$

and they observed that this gives elementary solutions of the spherical compaction problem for the discrete set of values

$$\varepsilon_c = 2n/3 \quad \text{or} \quad 2/3n \quad ; \quad (n=1,2,\ldots) \quad . \tag{22}$$

The particular value $\varepsilon_c = 2/3$ gives the empirical equation (19), with

$$A = \frac{2}{3} (Y_\infty - Y_0)\ln \frac{1}{1-D_\infty} \quad ; \quad B = \frac{2}{3} Y_\infty \quad . \tag{23}$$

(3) Rate effects

Murray, Rodgers and Williams [19] proposed a rate dependent compaction equation

$$P = C\frac{\dot{D}}{1-D} \quad ; \quad C = \frac{4}{3} \eta \quad , \tag{24}$$

where η is the shear viscosity and the dot denotes time derivative. This equation was adapted from an equation for pressureless sintering obtained by Mackenzie and Shuttleworth [11] from a hollow sphere model with linearly viscous fluid response.

Wilkinson and Ashby [20] developed a rate dependent compaction model from micromechanical analyses with power law creep response. Their model describes three stages of compaction, each stage being described by an equation of the form

$$\dot{D} = f(D)P^n \quad . \tag{25}$$

Modifications of this model were suggested by Swinkels et. al. [21].

Carroll and Holt [22] obtained a rate dependent compaction equation from a dynamic elastic-perfectly plastic hollow sphere model analysis. This equation was subsequently modified by Holt, Carroll and Butcher [23] to include a viscous rate effect, leading to an equation of the form

$$P = P_{st}(D) + P_{vis}(D,\dot{D}) + P_{kin}(D,\dot{D},\ddot{D}) \ . \tag{26}$$

The static response function $P_{st}(\ )$ describes an initial elastic phase, an elastic plastic phase and a fully plastic phase (described by equation (20)). The viscous response function $P_{vis}(\ )$ has the form of equation (24). The kinetic term $P_{kin}(\ )$ describes the inertial effect. The kinetic term may be written as

$$P_{kin} = -\kappa^2 \frac{d}{d\alpha} \{\tfrac{1}{2} \ h(\alpha)\dot{\alpha}^2\} \quad ; \quad \alpha = 1/D \quad , \tag{27}$$

with

$$\kappa^2 = \frac{\rho a_o^2}{3(\alpha_o-1)^{2/3}} \quad ; \quad h(\alpha) = (\alpha-1)^{-1/3} - \alpha^{-1/3} \ . \tag{28}$$

By multiplying equation (26) on both sides by $\dot{D}/D^2$ we obtain the energy balance equation for the compacting hollow sphere. The rate of work by the applied pressure equals the sum of the rates of elastoplastic work, viscous dissipation and kinetic energy. It is evident from equation (28) that the kinetic energy is proportional to the material density ρ and to the surface area of the pore. An equation of the form (26) may also be useful in describing wave propagation in bubbly fluids.

One can also obtain a simple rate dependent compaction equation of the form

$$P = -A + B\beta + C\dot{\beta} \tag{29}$$

with constants A, B and C given by equations (23) and (24), from a viscoplastic model with linear viscosity and exponential saturation hardening.

The new densification measure

$$\beta = \ln \frac{1}{1-D} \tag{30}$$

is introduced to simplify equation (29).

(4) <u>Pressure maximum</u>

An interesting effect occurs in spherical inflation under internal pressure (or external allround tension), namely the pressure may attain a maximum value and then decrease, or it may attain local maximum and minimum values, as the sphere continues to expand. This effect may be observed by experimentally in inflating a rubber balloon. It may also have important implications with regard to growth of voids in tensile stress fields, leading to rupture or fracture.

To study this effect, we write the material response law for uniaxial compressive true stress σ as

$$\sigma = \hat{\sigma}(\lambda) = \overline{\sigma}(\epsilon) \quad ; \quad (\epsilon = - \ln \lambda) , \tag{31}$$

where ϵ denotes the axial logarithmic compressive strain. The internal pressure (or allround tension) P is again given by equation (17). It is convenient to introduce a change of variable

$$x = \lambda^{3/2} = r_0^3/r^3 , \tag{32}$$

leading to

$$P = \frac{2}{3} \int_{x_a}^{x_b} \hat{\sigma}(x^{2/3}) \frac{dx}{1-x} , \tag{33}$$

with x_a and x_b given by

$$x_a = a_0^3/a^3 = (\alpha_0-1)/(\alpha-1) \quad ; \quad x_b = b_0^3/b^3 = \alpha_0/\alpha . \tag{34}$$

Differentiation of equation (33), using Leibnitz' rule, gives

$$\frac{dP}{d\alpha} = \frac{2}{3(\alpha-\alpha_0)} \left[\frac{\alpha_0-1}{\alpha-1} \, \hat{\sigma}\{(\frac{\alpha_0-1}{\alpha-1})^{2/3}\} - \frac{\alpha_0}{\alpha} \, \hat{\sigma}\{(\frac{\alpha_0}{\alpha})^{2/3}\}\right] . \tag{35}$$

It is convenient to introduce a response function q, defined by

$$q(x) = x\hat{\sigma}(x^{2/3}) , \tag{36}$$

in terms of which equation (35) becomes

$$\frac{dP}{d\alpha} = \frac{2}{3(\alpha-\alpha_0)} \left[q(\frac{\alpha_0-1}{\alpha-1}) - q(\frac{\alpha_0}{\alpha})\right]. \tag{37}$$

The condition for a stationary value of the applied pressure P is

$$q(\frac{\alpha_0-1}{\alpha-1}) = q(\frac{\alpha_0}{\alpha}) . \tag{38}$$

This condition may also be written as

$$\lambda_a^{3/2}\sigma_a = \lambda_b^{3/2}\sigma_b , \tag{39}$$

where λ_a , λ_b and σ_a , σ_b denote the values of the radial stretch λ and the compressive radial stress $\sigma = T_{\theta\theta} - T_{rr}$ at $r=a$ and at $r=b$. Differentiation of equation (37) and use of equation (38) gives an expression for $d^2P/d\alpha^2$ at a stationary point $\alpha = \alpha^*$:

$$\left(\frac{d^2P}{d\alpha^2}\right)^* = \frac{2}{3(\alpha-\alpha_0)} \left\{ -\frac{\alpha_0-1}{(\alpha-1)^2} \, q'(\frac{\alpha_0-1}{\alpha-1}) + \frac{\alpha_0}{\alpha^2} \, q'(\frac{\alpha_0}{\alpha})\right\} . \tag{40}$$

It follows from equations (38) and (40) that the qualitative behavior of the pressure P in spherical inflation is determined by the form of the function q, defined in equation (36), on the interval $(0,1)$. Monotonic uniaxial compressive stress response allows for different types of qualitative response in spherical inflation, the most important being Types A, B and C described below (Carroll [24]):

Type A: The function q is monotonic on $(0,1)$. Equation (38) does not have a real root in $\alpha_0 < \alpha < \infty$, so the pressure P increases monotonically.

Type B: The function q has a maximum value on $(0,1)$. Equation (38) has one real root α^* , with $\alpha_0 < \alpha^* < \infty$. The pressure P increases to a maximum value P^* at porosity α^* and then decreases.

Type C: The function q has a local maximum and a local minimum in $(0,1)$. Equation (38) may have no root, one root or two roots in (α_0, ∞) , depending on the initial porosity α_0 . The pressure P increases monotonically for thick walled spheres, but has a local maximum and minimum for thin walled spheres.

We remark that q may be expressed in terms of the logarithmic compressive strain ϵ as follows:

$$q = \lambda^{3/2}\hat{\sigma}(\lambda) = e^{-3\epsilon/2}\bar{\sigma}(\epsilon) . \tag{41}$$

The condition that the function q be monotonic is simply that

$$\frac{d\sigma}{d\epsilon} > \frac{3}{2}\sigma . \tag{42}$$

The behavior in spherical inflation is of Type A if the condition (42) is met for all strains, it is of Type B if the condition is met only for a finite portion $0 < \epsilon < \epsilon_1$ and it is of Type C if the condition is met for all strains ϵ with $0 < \epsilon < \epsilon_1$ or $\epsilon_2 < \epsilon < \infty$.

The foregoing analysis applies equally to nonlinearly elastic materials and to elastic-plastic materials. The behavior of metals is usually of Type B. It is interesting to observe that the important class of Mooney-Rivlin elastic solids, which have a strain-energy function

$$W = C_1(I_1-3) + C_2(I_2-3) , \tag{43}$$

with C_1 and C_2 constant and I_1 and I_2 the principal strain invariants, admits all three types of qualitative behavior. The behavior is of Type A for $C_2/C_1 > \kappa_{cr}$, of Type C for $0 < C_2/C_1 < \kappa_{cr}$, and of Type B for neoHookean materials ($C_2=0$), where κ_{cr} is given approximately by $\kappa_{cr} \simeq 0.2145$.

(5) <u>Deviatoric effects</u>

The hollow sphere model can also provide some insight into the nonhydrostatic response of porous materials. As mentioned previously, linearly elastic solutions with stress free inner boundary and with homogeneous tractions on the outer boundary, i.e.,

$$t_i = \sigma^o_{ij} n_j \quad \text{on} \quad r = b \quad ; \quad (\sigma^o_{ij} \text{ constants}) , \tag{44}$$

give effective moduli for poroelastic solids (Mackenzie [12]). Study of these solutions also gives an initial yield condition, i.e., a surface in σ^o_{ij}-space for which the loading (44) gives rise to yielding at the inner boundary of the sphere (Curran and Carroll [25]). Beyond the elastic range, one must resort to approximate solutions or numerical methods. Green [26] and Gurson [27] have presented such solutions for elastic-perfectly plastic solids, giving yield conditions

$$\frac{1}{6} \left\{ \frac{3-2\phi^{1/4}}{1-\phi^{1/3}} \right\}^2 s_{ij}s_{ij} + \left\{ \frac{\sigma_{kk}}{2 \ln \phi} \right\}^2 = Y^2 \tag{45}$$

and

$$\frac{3}{2} s_{ij}s_{ij} + 2\phi Y^2 \cosh\left(\frac{\sigma_{kk}}{2Y}\right) = (1+\phi^2)Y^2 , \tag{46}$$

where s_{ij} are deviatoric stress components. Curran and Carroll [25] also used finite element solutions to exhibit shear enhanced compaction.

More recently, Carman and Carroll [28] have used a finite element code FEAP, with finite strain elastic-plastic response laws, in the hollow sphere model. The results tend to support the assumption that the yield sufrace is a function of stress and porosity only. Results for elastic-perfectly plastic materials show good agreement with the yield conditions of Green and Gurson. An advantage of this model is that one can also run strain controlled tests, of the form

$$x_i = \lambda_i X_i \quad (\text{no sum}) \quad \text{on} \quad r = b , \tag{47}$$

where x_i and X_A denote rectangular Cartesian coordinates in the deformed state and in the reference state. In particular the (average) stress path in a constant volume test $(\lambda_1 \lambda_2 \lambda_3 = 1)$ gives the yield surface at any particular value of the porosity. We are presently using this model to examine the effect of strain hardening on the yield surfaces. We are also examining the intriguing question of local maxima (or loss of monotonicity) in nonhyrostatic loading, by considering loading paths of the form

$$\varepsilon_i = \ell n\ \lambda_i = \varepsilon m_i \quad (m_i m_i = 1) , \tag{48}$$

i.e., rectilinear paths in principal logarithmic strain space.

4. Creep Compaction of Metal Powders

We now describe a model for time dependent densification of metal powders at constant pressure and temperature (Carroll [29]). The model is semi-empirical; it is based in part on micromechanical analysis of the hollow sphere model and in part on experimental data of Swinkels et. al. [21].

(1) The experiments

The data was obtained from an excellent set of experiments on hot isostatic pressing of lead and tin powders [21]. In a typical experiment, the powder was heated to a particular temperature, which was then held constant, and brought quickly (in 30 secs) to a particular pressure, which was then held constant. The rig allowed continuous measurement of density during the test and the creep densification was measured for 24 hrs. These tests were carried out for lead powder and for tin powder, for six different values of temperature (ranging from room temperature to 150°C) and usually for four or five different pressures at each temperature. The published data included the initial (30 secs) and final (24 hrs) relative density for each test and also four creep curves for each material at 100°C.

(2) <u>The</u> <u>model</u>

The data on initial densification shows that the value of the relative density after 30 seconds depends significantly on the pressure level. This suggests that a creep response law will not provide an adequate description. (If the rise time is small compared with the characteristic time of the material, then the change in relative density during the pressure rise will also be small.) The data thus suggests a structure

$$\frac{d}{dt} \{P-P_i(\beta)\} = F\{P-P_e(\beta)\} \, , \tag{49}$$

where the functions $P_i(\ )$ and $P_e(\ )$ describe instantaneous and equilibrium response laws and the function $F(\ )$ describes creep response. This equation describes other processes in addition to that carried out by Swinkels et. al. [21], for example, stress relaxation at constant relative density.

The data in [21] was presented in terms of pressure, temperature and relative density D. The widespread use of the Konopicky equation, and the discussion of rate effects in the previous section, suggest that the densification β would be a more convenient choice of kinematic variable. We assumed that both the instantaneous and equilibrium response laws are of the Konopicky-Shapiro and Kolthoff type, so that

$$P = -A_i + B_i\beta(0) \tag{50}$$

and

$$P = -A_e + B_e\beta_\infty \, , \tag{51}$$

where $\beta(0)$ and β_∞ denote the initial and final densifications. The experiments of Swinkels et.al. provide four points for initial densification and four for final densification at 100°C for each material and so they provide a check on the assumptions (50) and (51). It is evident from Figure 1 that these assumptions are very good ones for both lead and tin. Indeed, it is interesting to observe that the constants A_i and B_i have the same value for both materials. The assumptions (50) and (51) hold at the other temperatures also.

The form of the creep response law was found by plotting densification rate $\dot{\beta}$ versus residual densification $\beta_\infty - \beta$. This suggested a creep law of the form

$$\tau\dot{\beta} = e^{m(\beta_\infty - \beta)} - 1 . \tag{52}$$

The asymptotic behavior is a linear creep law, and comparison with equation (29) affords the identification

$$\tau/m = 2\eta/Y_\infty . \tag{53}$$

Thus, while the exponential creep law (52) is empirical, micromechanical analysis gives the ratio of the constants τ and m.

Integration of equation (52), with zero time densification $\beta(0)$, gives

$$1 - e^{-m(\beta_\infty - \beta)} = \{1 - e^{-m(\beta_\infty - \beta(0))}\}e^{-mt/\tau} . \tag{54}$$

The corresponding expression for the relative density is of the form

$$1 - D = \frac{1 - D_\infty}{(1 - Ke^{-mt/\tau})^{1/m}} ; \quad K = 1 - \{\frac{1 - D_\infty}{1 - D(0)}\}^m . \tag{55}$$

This model gave an excellent fit with experimental data for hot isostatic pressing of lead and tin powders [22]. The theoretical creep curves and experimental data for tin powder at 100°C are shown in Figure 2.

Acknowledgments

This work was supported by a contribution from the Shell Companies Foundation, in support of a Shell Distinguished Chair at the University of California, Berkeley, and also by Grant. No. MEA-820534 from the National Science Foundation (Solid Mechanics Program) to the University of California, Berkeley. I am grateful for this support.

References

1. W. Herrmann, J. Appl. Phys., $\underline{40}$, 2490 (1969).

2. M.A. Goodwin and S.C. Cowin, Arch. Rat. Mech. Anal., $\underline{44}$, 249 (1972).

3. R.B. Martin, Pore Structure and Properties of Materials, $\underline{1}$, A35 (edited by S. Modry), Academia, Prague (1973).

4. D.A. Drew, Studies in Applied Math. $\underline{50}$, 133 and 205 (1971).

5. Mechanics of Granular Materials. New Models and Constitutive Relations (edited by J.T. Jenkins and M. Satake), Elsevier, Amsterdam, Oxford, New York (1983).

6. The Effect of Voids on Material Deformation (edited by S.C. Cowin and M.M. Carroll), AMD Vol. 16, ASME, New York (1976).

7. Journal of the Engineering Mechanics Division, ASCE $\underline{106}$ (1980).

8. International Journal of Engineering Science, $\underline{22}$ (1984)

9. M.M. Carroll and N. Katsube, J. Energy Res. Tech. $\underline{105}$, 509 (1983).

10. C. Torre, Berg-Huttenmann. Monatsh. Montan. Hochschule Leoben $\underline{93}$, 62 (1948).

11. J.K. Mackenzie and R. Shuttleworth, Proc. Phys. Soc. $\underline{1362}$, 833 (1949).

12. J.K. MacKenzie, Proc. Phys. Soc. $\underline{B62}$, 2 (1950).

13. Z. Hashin, J. Appl. Mech. $\underline{29}$, 143 (1962).

14. K. Knopicky, Radex Rundschau $\underline{3}$, 141 (1948).

15. I. Shapiro and I.M. Kolthoff, J. Phys. Colloid. Chem. $\underline{51}$, 483 (1947).

16. M.M. Carroll and K.T. Kim, Powder Metallurgy $\underline{29}$, 153 (1984).

17. E. Voce, Metallurgica $\underline{51}$, 219 (1955).

18. J.H. Palm, Appl. Scient. Res. A2, 198 (1949).

19. P. Murray, E.P. Rodgers and J. Williams, Trans. Br. Ceram. Soc. $\underline{53}$, 474 (1954).

20. D.S. Wilkinson and M.F. Ashby, <u>Proc. 4th Int. Conf. on Sintering and Catalysis</u>, 472, Pergamon Press, New York (1975).

21. F.B. Swinkels, D.S. Wilkinson, E. Arzt and M.F. Ashby, Acta Metall. $\underline{31}$, 1829 (1983).

22. M.M. Carroll and A.C. Holt, J. Appl. Phys. $\underline{72}$, 1326 (1972).

23. A.C. Holt, M.M. Carroll and B.M. Butcher, _Pore Structure and Properties of Materials_, 5, D63 (edited by S. Modry), Academia, Prague (1974).

24. M.M. Carroll, submitted for publication.

25. J.H. Curran and M.M. Carroll, J. Geophys. Res. 84, 1105 (1979).

26. R.J. Green, Int. J. Mech. Sci. 14, 215 (1972).

27. A.L. Gurson, Trans. ASME J. of Engng. Mats. & Technology 99, 2 (1977).

28. R. Carman and M.M. Carroll, to appear in Mechanics of Materials.

29. M.M. Carroll, submitted for publication.

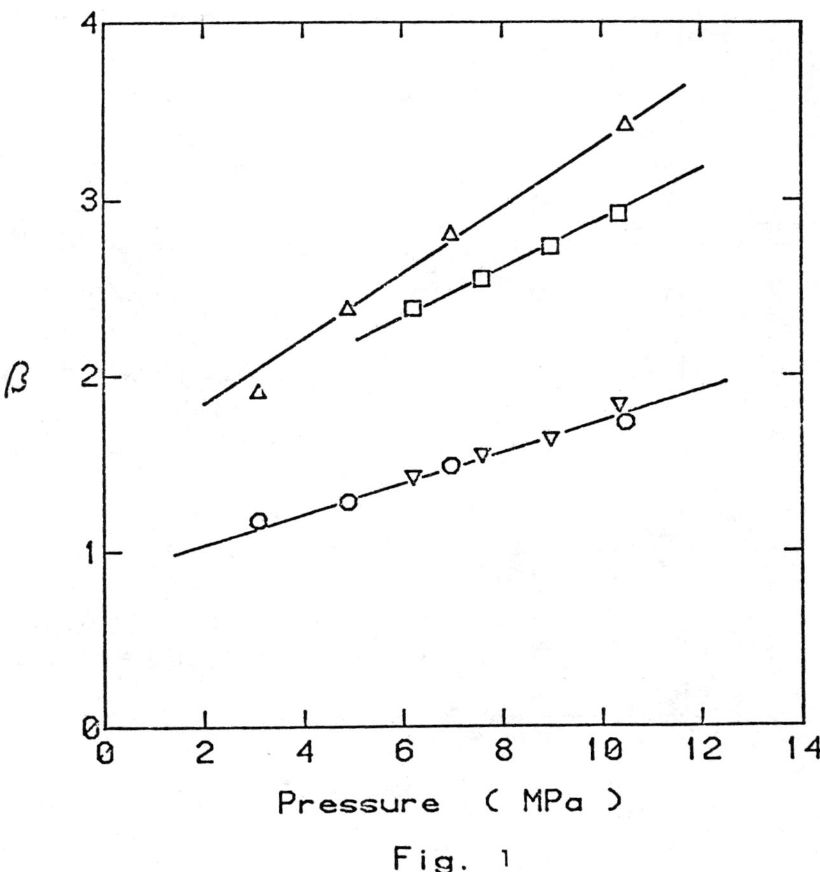

Figure 1.   Pressures, zero time densifications and final densifications for pressure sintering of lead (O and □) and tin (∇ and Δ), from [21].

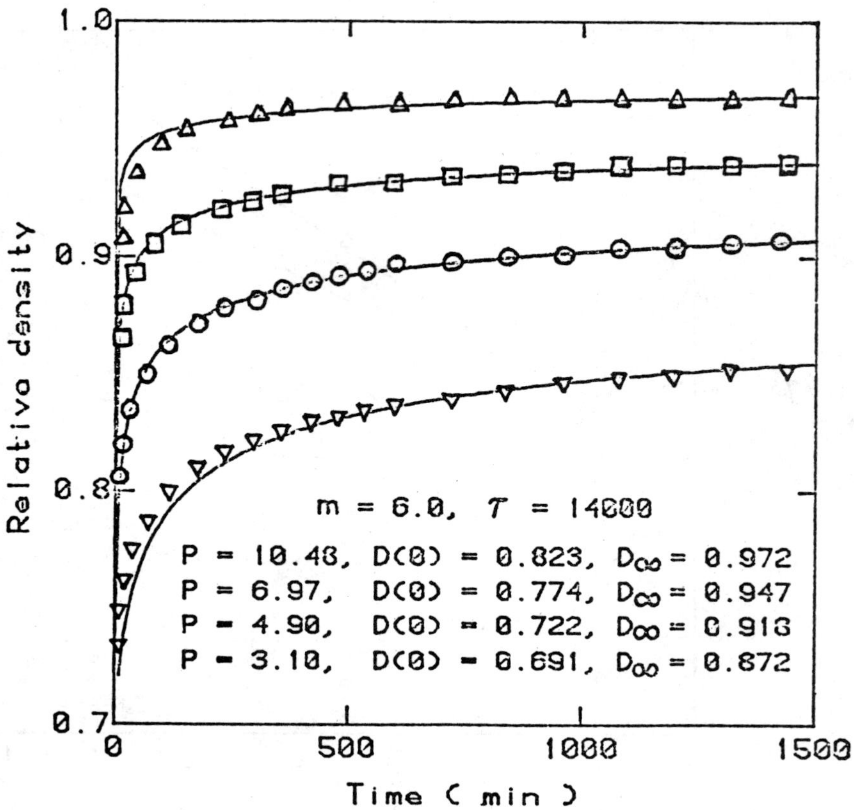

Figure 2   Comparison of theoretical creep compaction curves (——) for tin at
100°C, from equation (24), with experimental data from [21].

ON BOUNDING THE EFFECIVE CONDUCTIVITY
OF ANISOTROPIC COMPOSITES

Robert V. Kohn

Courant Institute
251 Mercer Street
New York, NY 10012

Graeme W. Milton

Department of Physics
California Institute of Technology
Pasadena, CA 91125

1. Introduction

There has recently been a renewal of interest in bounding the effective moduli of composite materials. Several factors are responsible, including attention to applications in structural optimization [1,18,20,28]. The developments of the past several years include new ways of applying old variational principles such as those of Hashin and Shtrikman. They also include two entirely new methods for proving bounds: one based on compensated compactness, and the other on explicit representation formulas.

One result of this activity has been the specification of optimal bounds for anisotropic composites made by mixing two isotropic conducting materials in specified proportions. First established using the compensated compactness method [21,22,33], the bounds have also been proved using representation formulas [24], and a third proof is possible using the Hashin-Shtrikman variational principle [17]. Their applications to structural optimization are discussed in [19] and [28].

This paper presents, in self-contained form, each of the three approaches to proving these bounds. Our goal is to popularize the methods, and to highlight their similarities and differences, by explaining them in the context of this simple but fundamental example.

2. The Bounds

From the physical viewpoint, we are interested in composites made by mixing two isotropic materials with conductivities α and β, $0 < \alpha < \beta$, in specified proportions θ_α and $\theta_\beta = 1 - \theta_\alpha$. The effective conductivity A is a symmetric second-order tensor relating the (local) average electric field to the (local) average current flow. In the case of a spatially periodic composite the averaging is over a period cell [2,30]; for a random one it is over the statistical ensemble [29]. There is also a third, more general approach, which replaces averaging by weak convergence [6,25,38]. For simplicity, we shall only discuss the <u>spatially periodic</u> case. This represents no loss of generality, since the effective conductivity of any composite can be approximated by spatially periodic ones with increasingly complicated microstructures. (This is established for random composites in [10]; a proof using the theory of G-convergence will appear in [16].)

Mathematically, then, our composites will be described on the microscopic scale by a periodic function $a(y)$, with period cell $Q \subset R^n$, taking only the values α and β. The effective conductivity A is defined by the variational principle

$$\langle A\xi, \xi \rangle = \inf_\phi \fint_Q a|\xi + \nabla\phi|^2 dy$$

for any $\xi \in R^n$, where ϕ ranges over Q-periodic functions and $\fint_Q$ devotes the average over Q. If $\chi_\alpha(y)$ and $\chi_\beta(y)$ denote the characteristic functions of the sets where a equals α and β, respectively, then

$$a(y) = \alpha\, \chi_\alpha + \beta\, \chi_\beta$$

and the volume fractions are

$$\theta_\alpha = \fint_Q \chi_\alpha\, dy \quad , \quad \theta_\beta = \fint_Q \chi_\beta\, dy \ .$$

An elementary bound on A has been known early in this century:

$$\text{(EB)} \qquad hI \leq A \leq mI \quad ,$$

where h and m denote the harmonic and arithmetic means of A, respectively,

$$h = (\textstyle\int_0 a^{-1} dy)^{-1} = (\alpha^{-1}\theta_\alpha + \beta^{-1}\theta_\beta)^{-1}$$

$$m = \textstyle\int_0 a\, dy = \alpha\theta_\alpha + \beta\theta_\beta \; .$$

This bound is _sharp_, in the sense that for a laminar microstructure the largest
eigenvalue of A equals h. However, it is not _optimal_: not every matrix A
which satisifes (EB) is the effective conductivity of a composite made by mixing
α and β in proportions θ_α and θ_β .

The _optimal_ _bounds_ were first established by Murat and Tartar [33] and by
Lurie and Cherkaev [21,22]. In addition to (EB), A must satisfy the lower bound

(LB)
$$\mathrm{tr}[(A - \alpha I)^{-1}] \; \leq \; \frac{n-1}{m-\alpha} \; + \frac{1}{h-\alpha}$$

and the upper bound

(UB)
$$\mathrm{tr}[(\beta I - A)^{-1}] \; \leq \; \frac{n-1}{\beta-m} \; + \frac{1}{\beta-h} \; .$$

If A satisfies these three bounds, then a composite with effective conductivity
A and volume fractions θ_α, θ_β can be constructed using coated ellipsoids, a
generalization of the well-known coated sphere construction that works in the
isotropic case [33]. There is also a second, entirely different construction
using successive lamination [21,22,33] (c.f. the contributions of Milton and
Tartar). The details of these constructons will not be repeated here; our
attention is focused instead on the proofs of (LB) and (UB).

Some remarks are in order about the relation of these results to earlier
work. Bounds equivalent to (LB) and (UB) were established by Hashin and
Shtrikman in 1962, for the special case of _random, statistically isotropic_ composites [13]. A mathematically more rigorous proof was subsequently given by
Willis [34]. An extensive literature has developed from this pioneering work; it
is reviewed, for example, in [4,12,35,36]. Much of it addresses the specification of better bounds for the effective conductivity A in terms of additional
statistical information, such as the two-point correlation function of the
(randomly varying) microscopic conductivity. The task of improving (EB) for
anisotropic composites, in terms of the volume fractions alone, seems simply to

have been ignored for many years. It was rediscovered in the 1970's by specialists in optimal control theory, who realized that optimal bounds such as (EB), (LB), (UB) determine the relaxation of a class of distributed parameter control problems.

3. Variational Principles

We have defined the effective conductivity by means of a variational principle

$$(3.1) \qquad \langle A\xi,\xi \rangle = \inf_{\phi} \; \int_Q a(y)|\xi + \nabla\phi|^2 dy.$$

The minimizer ϕ^ξ solves

$$(3.2) \qquad \nabla\cdot(a(y)(\xi + \nabla\phi^\xi)) = 0 \; ,$$

and integration by parts in (3.1) gives

$$(3.3) \qquad \langle A\xi,\xi \rangle = \int_Q \langle a(y)(\xi + \nabla\phi^\xi),\xi \rangle \, dy.$$

Since ϕ^ξ depends linearly on ξ , it follows by polarization that

$$(3.4) \qquad A\xi = \int_Q a(y)(\xi + \nabla\phi^\xi) dy.$$

3A. <u>Dual variational principles.</u>

A second characterization of A is obtained by taking the convex dual of (3.1). We have

$$\frac{1}{2} \langle E\xi,\xi \rangle \; \geqslant \; \langle \xi,\sigma \rangle - \frac{1}{2} \langle E^{-1}\sigma,\sigma \rangle$$

for any ξ, σ R^n and any positive, symmetric matrix E, with equality precisely when $\sigma = E\xi$. Therefore (3.1) can be written

$$(3.5) \qquad \frac{1}{2} \langle A\xi,\xi \rangle = \inf_{\phi} \sup_{\sigma} \int_Q [\langle \xi + \nabla\phi,\sigma \rangle - \frac{1}{2} a(y)^{-1}|\sigma|^2] dy \; ,$$

in which σ ranges over Q-periodic vector fields. Interchanging the inf and sup in (3.5) yields

$$(3.6) \qquad \frac{1}{2} \langle A\xi,\xi \rangle = \sup_{\sigma} \inf_{\phi} \int_Q [\langle \xi + \nabla\phi, \sigma \rangle - \frac{1}{2} a(y)^{-1} |\sigma|^2] dy$$

If $\operatorname{div} \sigma \neq 0$ then the infimum over ϕ in (3.6) is $-\infty$, not of interest since we are maximizing over σ. If $\operatorname{div} \sigma = 0$ then $\int_Q \langle \nabla\phi, \sigma \rangle dy = 0$ and so

$$(3.7) \qquad \frac{1}{2} \langle A\xi,\xi \rangle = \sup_{\operatorname{div} \sigma = 0} \int_Q [\langle \xi, \sigma \rangle - \frac{1}{2} a(y)^{-1} |\sigma|^2] dy.$$

This is the variational principle dual to (3.1).

A word about the passage from (3.5) to (3.6). The fact that inf sup = sup inf follows from a general saddle-point theorem [7, Prop. 2.2, page 173]. For quadratic variational problems there is also an easy, direct proof of duality: for any periodic σ with $\operatorname{div} \sigma = 0$,

$$\frac{1}{2} \langle A\xi,\xi \rangle = \int_Q \frac{1}{2} \langle a(y)(\xi + \nabla\phi^\xi), (\xi + \nabla\phi^\xi) \rangle dy$$

$$(3.8) \qquad \geq \int_Q [\langle \sigma, \xi + \nabla\phi^\xi \rangle - \frac{1}{2} \langle a(y)^{-1}\sigma, \sigma \rangle] dy$$

$$= \int_Q [\langle \sigma, \xi \rangle - \frac{1}{2} \langle a(y)^{-1}\sigma, \sigma \rangle] dy$$

with equality in (3.8) when σ is

$$(3.9) \qquad \sigma^\xi = a(y)(\xi + \nabla\phi^\xi).$$

There is a modification of the dual principle (3.7) which is more symmetric to (3.1):

$$(3.10) \qquad \langle A^{-1}\overline{\sigma}, \overline{\sigma} \rangle = \inf_{\substack{\operatorname{div} \sigma = 0 \\ \int\sigma = \overline{\sigma}}} \int_Q a(y)^{-1} |\sigma|^2 \, dy,$$

for any $\overline{\sigma} \in R^n$. To prove it, we observe that the extremal σ^ξ of (3.7) has mean value

$$\overline{\sigma}^\xi = \int_Q \sigma^\xi dy = \int_Q a(y)(\xi + \nabla\phi^\xi) dy = A\xi.$$

This can therefore be imposed as a constraint in (3.7) without changing the value of the supremum. Hence

$$\frac{1}{2} \langle A\xi,\xi \rangle = \sup_{\substack{\text{div}\,\sigma=0 \\ f\sigma \,=\, A\xi}} f_Q [\langle \xi,\sigma \rangle - \frac{1}{2} a(y)^{-1} |\sigma|^2] dy$$

$$= \langle A\xi,\xi \rangle - \frac{1}{2} \inf_{\substack{\text{div}\,\sigma=0 \\ f\sigma=A\xi}} f_Q a(y)^{-1} |\sigma|^2 dy.$$

Rearrangement and the substitution $\overline{\sigma} = A\xi$ leads to (3.10).

The elementary bounds (EB) are an immediate consequence of these variational principles: The choice $\phi=0$ in (3.1) gives $A < mI$, while $\sigma \equiv \overline{\sigma}$ in (3.10) gives $A > hI$.

3B. Hashin-Shtrikman variational principles.

These principles were introduced by Hashin and Shtrikman in [13], and clarified by Hill [14] and Willis [34]. The key idea is to dualize not the full energy $a(y) |\xi + \nabla\phi|^2$, but only a part of it. We shall present two versions, one giving a lower bound and the other an upper bound for $\langle A\xi,\xi \rangle$.

We begin by adding and subtracting $\gamma |\xi + \nabla\phi|^2$ — the energy of a "reference medium" with constant conductivity γ — on the right of (3.1):

$$(3.11) \qquad \frac{1}{2} \langle A\xi,\xi \rangle = \inf_{\phi} \frac{1}{2} f_Q [(a(y)-\gamma) |\xi + \nabla\phi|^2 + \gamma |\xi + \nabla\phi|^2] dy.$$

If γ is restricted to the range $0 < \gamma < \alpha$, then $a(y) - \gamma > 0$, and the first term on the right can be written as a supremum over Q - periodic vector fields σ:

$$(3.12) \quad \frac{1}{2} \langle A\xi,\xi \rangle = \inf_{\phi} \sup_{\sigma} f_Q [\langle \sigma, \xi + \nabla\phi \rangle - \frac{1}{2} (a(y)-\gamma)^{-1} |\sigma|^2 + \frac{\gamma}{2} |\xi + \nabla\phi|^2] dy.$$

We interchange the inf and sup - appealing again to convex analysis [7] or arguing as section 3A - and do the minimization on ϕ directly. The optimal choice solves

$$(3.13) \qquad \Delta\phi = -\frac{1}{\gamma} \ \text{div} \ \sigma \ ;$$

it is admissible, because the Laplacian is an isomorphism from periodic H^1 functions to periodic H^{-1} functions with mean value zero. Substitution in (3.12) and integration by parts gives

$$(3.14) \ \tfrac{1}{2} \langle(A-\gamma)\xi,\xi\rangle = \sup_{\sigma} \int_0 [\langle\sigma,\xi\rangle - \tfrac{1}{2} (a(y)-\gamma)^{-1}|\sigma|^2 - \frac{1}{2\gamma} \langle\sigma,\nabla\Delta^{-1}\text{div} \ \sigma\rangle] dy.$$

This is the variational principle of Hashin and Shtrikman for bounding A below.

For the second principle, we return to (3.11). If γ is restricted to the range $\beta < \gamma < \infty$, then $a(y) - \gamma$ is negative and the first term on the right becomes an infimum:

$$(3.15) \qquad \tfrac{1}{2} \langle A\xi,\xi\rangle = \inf_{\phi} \inf_{\sigma} \int_0 [\langle\sigma,\xi + \nabla\phi\rangle - \tfrac{1}{2} (a(y)-\gamma)^{-1}|\sigma|^2 + \tfrac{\gamma}{2} |\xi+\nabla\phi|^2] dy.$$

We interchange the order of minimization - no convex duality is required here - and evaluate the infimum over ϕ. The optimal choice is again (3.13), and substitution into (3.15) gives

$$(3.16) \ \tfrac{1}{2} \langle(A-\gamma)\xi,\xi\rangle = \inf_{\sigma} \int_0 [\langle\sigma,\xi\rangle - \tfrac{1}{2} (a(y)-\gamma)^{-1}|\sigma|^2 - \frac{1}{2\gamma} \langle\sigma,\nabla\Delta^{-1}\text{div} \ \sigma\rangle] dy.$$

This is the Hashin-Shtrikman variational principle for bounding A above. Note that the integrands in (3.14) and (3.16) are the same, except that γ must be less than α for (3.14) and greater than β for (3.16).

4. Bounds via HS Variational Principles

We shall make the customary substitution of a piecewise constant σ into the Hashin-Shtrikman variational principle. This leads to a bound for the effective conductivity A in terms of a tensor F of "geometric parameters" defined by

$$(4.1) \qquad \langle Fn,n\rangle = \int_0 \langle n\chi, \nabla\Delta^{-1}\text{div}(n\chi) \rangle \ dy,$$

where $n \in R^n$ and χ is either χ_α or χ_β , the characterisic functions of the two materials. The traditional approach for random composites is to write the expected value of F in terms of the kernel for $(\nabla \Delta^{-1} \text{div})$ and the two-point correlation function of the random field χ. With a hypothesis of statistical isotropy, Hashin and Shtrikman used the symmetries of the kernel to evaluate the resulting expression in terms of the volume fractions alone. We shall proceed differently, by deriving a relation on the _trace_ of F irrespective of the symmetry of the composite. Algebraic manipulation then leads to the desired bound.

4A. Constraints on the geometric parameters F_{ij}.

The key to our approach is this simple result.

Lemma 4.1: _Let_ χ _be any_ Q - _periodic characteristic function with mean value_ $\bar{\chi} = \int_Q \chi dy$. _Then the matrix_ F _defined by_ (4.1) _is symmetric and nonnegative, with trace_ $\bar{\chi}(1-\bar{\chi})$.

Proof: Since $\chi - \bar{\chi}$ has mean value zero, there is a periodic solution of

$$\Delta \psi = \chi - \bar{\chi} .$$

Differentiating the equation gives

$$\frac{\partial \psi}{\partial y_i} = \Delta^{-1} \frac{\partial \chi}{\partial y_i} , \qquad 1 < i < n,$$

and so

$$(4.2) \qquad \frac{\partial}{\partial y_i} \Delta^{-1} \frac{\partial}{\partial y_j} \chi = \frac{\partial^2 \psi}{\partial y_i \partial y_j}.$$

By definition

$$F_{ij} = \int_Q \chi \frac{\partial}{\partial y_i} \Delta^{-1} \frac{\partial}{\partial y_j} \chi \, dy,$$

and substitution of (4.2) into this gives

$$F_{ij} = \int_Q \chi \frac{\partial^2 \psi}{\partial y_i \partial y_j} dy = \int_Q (\chi - \bar{\chi}) \frac{\partial^2 \psi}{\partial y_i \partial y_j} dy$$

$$(4.3) \qquad = \int_Q \Delta \psi \cdot \frac{\partial^2 \psi}{\partial y_i \partial y_j} dy$$

$$(4.4) \qquad = \oint_Q \langle \nabla \frac{\partial \psi}{\partial y_i}, \nabla \frac{\partial \psi}{\partial y_j} \rangle \, dy.$$

Symmetry is clear from (4.3), and positivity is a consequence of (4.4), since it yields

$$\langle Fn, n \rangle = \oint_Q |\nabla(n \cdot \nabla \psi)|^2 dy \qquad\qquad \forall n \quad \mathbb{R}^n.$$

The trace is easily computed from (4.3):

$$\mathrm{tr}\, F = \oint_Q \Delta \psi \cdot \Delta \psi \, dy = \oint_Q (x - \overline{x})^2 dy = \overline{x}(1 - \overline{x}).$$

4B. The lower bound.

We begin with the Hashin-Shtrikman lower bound (3.14). The test field σ is chosen to be a constant vector n where $a(y) = \beta$, and zero elsewhere:

$$\sigma = n \chi_\beta \, .$$

Substitution into (3.14) and passage to the limit $\gamma \uparrow \alpha$ gives

$$(4.5) \qquad \frac{1}{2} \langle (A - \alpha I)\xi, \xi \rangle - \theta_\beta \langle n, \xi \rangle \; \geq \; - \frac{1}{2} \theta_\beta (\beta - \alpha)^{-1} |n|^2 - \frac{1}{2\alpha} \langle F^\beta n, n \rangle,$$

where F^β is defined by (4.1) with $\chi = \chi_\beta$.

The relation (4.5) holds for all vectors $\xi, n \in E^n$. Fixing n, we choose ξ to minimize the left hand side:

$$\xi = \theta_\beta (A - \alpha I)^{-1} n.$$

Substitution into (4.5) gives

$$\frac{1}{2} \theta_\beta^2 \langle (A - \alpha I)^{-1} n, n \rangle \leq \frac{\theta_\beta}{2(\beta - \alpha)} |n|^2 + \frac{1}{2\alpha} \langle F^\beta n, n \rangle.$$

This is a relation between symmetric matrices,

$$(A - \alpha I)^{-1} \leq \frac{1}{\theta_\beta (\beta - \alpha)} I + \frac{1}{\alpha \theta_\beta^2} F^\beta \, .$$

Taking the trace and using Lemma 4.1, we conclude that

$$(4.6) \qquad \mathrm{tr}[(A - \alpha I)^{-1}] < \frac{n}{\theta_\beta(\beta-\alpha)} + \frac{(1-\theta_\beta)}{\alpha\theta_\beta} .$$

The right hand side of (4.6) is easily seen to agree with that of (LB).

4C. The upper bound.

This time we substitute

$$\sigma = \eta \, \chi_\alpha$$

into the Hashin-Shtrikman upper bound (3.16). After passage to the limit $\gamma\!\downarrow\!\beta$ this yields

$$(4.7) \qquad \tfrac{1}{2} \langle (A - \beta I)\xi,\xi \rangle - \theta_\alpha \langle \eta,\xi \rangle < \tfrac{1}{2} \theta_\alpha(\beta-\alpha)^{-1} - \frac{1}{2\beta} \langle F^\alpha \eta,\eta \rangle,$$

where F^α is defined by (4.1) with $\chi = \chi_\alpha$. We choose ξ to maximize the left hand side:

$$\xi = \theta_\alpha(A - \beta I)^{-1} \eta .$$

Substitution into (4.7) leads to

$$\tfrac{1}{2} \theta^2 \langle (\beta I - A)^{-1} \eta,\eta \rangle < \tfrac{1}{2} \theta_\alpha(\beta-\alpha)^{-1}|\eta|^2 - \frac{1}{2\beta} \langle F^\alpha \eta,\eta \rangle,$$

or in other words

$$(\beta I - A)^{-1} < \frac{1}{\theta_\alpha(\beta-\alpha)} \, I - \frac{1}{\beta\theta_\alpha^2} \, F^\alpha .$$

Taking the trace and applying Lemma 4.1, it follows that

$$(4.8) \qquad \mathrm{tr}[(\beta I - A)^{-1}] < \frac{n}{\theta_\alpha(\beta-\alpha)} - \frac{(1-\theta_\alpha)}{\beta\theta_\alpha} .$$

The right hand side of (4.8) is easily seen to agree with that of (UB).

5. Bounds via Compensated Compactness

The arguments in this section are drawn from [21,22,33]. Since the desired
bounds are relations among the eigenvalues of A, it is natural to use the basic
variational principles (3.1) or (3.10) simultaneously for n different choices
of ξ or $\bar{\sigma}$. This leads to variational principles for $\langle A\xi,\xi\rangle$ and $\langle A^{-1}\bar{\sigma},\bar{\sigma}\rangle$
when ξ and $\bar{\sigma}$ are $n \times n$ __matrices.__ Whereas the method of Hashin and Shtrikman
adds and subtracts the energy of a reference medium, the compensated compactness
method adds an indefinite quadratic form whose integral is controlled. The bound
is obtained by applying convex duality to the resulting expression.

5A. __More variational principles.__

We shall use $\langle\cdot,\cdot\rangle$ for the Hilbert-Schmidt inner product on $n \times n$ matrices

$$\langle X,Y\rangle = tr(X^t Y)$$

as well as for the inner product on R^n.

The basic variational principle (3.1) applied to n different vectors
$\xi^1,\ldots, \xi^n$ gives

$$\sum_{j=1}^{n} \langle A\xi^j, \xi^j\rangle = \inf_\phi \, f_0 \, a(y) \sum_{j=1}^{n} |\xi^j + \nabla\phi_j|^2 dy,$$

where $\phi = (\phi_1,\ldots, \phi_n)$ ranges over periodic, R^n-valued functions. Denoting by ξ
the n×n matrix with columns $\{\xi^j\}$, this can be written

$$(5.1) \qquad \langle A\xi,\xi\rangle = \inf \, f_0 \, a(y) \, |\xi+\nabla\phi|^2 dy.$$

It is identical to (3.1), except that ξ is now an n×n matrix and ϕ a vector-
valued function.

Similarly, applying (3.10) to n different vectors $\bar{\sigma}^1,\ldots, \bar{\sigma}^n$ yields

$$\sum_{j=1}^{n} \langle A^{-1}\sigma^j, \sigma^j\rangle = \inf_{\substack{div\,\sigma^j=0 \\ f_0\sigma^j = \bar{\sigma}^j}} \, f_0 \, a(y)^{-1} \sum_{j=1}^{n} |\sigma^j|^2 \, dy,$$

Writing $\bar{\sigma}$ for the $n \times n$ matrix with columns $\{\bar{\sigma}^j\}$, this becomes

$$(5.2) \qquad \langle A^{-1}\bar{\sigma},\bar{\sigma}\rangle = \inf_{\substack{\mathrm{div}\,\sigma=0 \\ \fint_Q \sigma = \bar{\sigma}}} \fint_Q a(y)^{-1} \sum |\sigma|^2 \, dy.$$

It is identical to (3.10), except that $\bar{\sigma}$ is an $n \times n$ matrix and σ a periodic, matrix-valued function whose columns are divergence-free.

5B. Special quadratic functions.

We shall specify two quadratic functions of $n \times n$ matrices, R and S, such that

$$(5.3) \qquad R(\xi) \leq \fint_Q R(\xi + \nabla\phi)\,dy$$

$$(5.4) \qquad S(\bar{\sigma}) \leq \fint_Q S(\sigma)\,dy$$

whenever ϕ and σ are admissible for (5.1) and (5.2). Any convex function would have these properties, but the useful examples are <u>nonconvex</u>. They are provided by the following two lemmas.

Lemmas 5.1: The choice

$$(5.5) \qquad R(\xi) = |\xi|^2 - (\mathrm{tr}\,\xi)^2$$

<u>satisfies</u> (5.3) <u>for any</u> $n \times n$ <u>matrix</u> ξ <u>and any periodic,</u> R^n-<u>valued,</u> H^1 <u>function</u> ϕ.

<u>Proof</u>: The difference between the right and the left sides of (5.3) equals

$$(5.6) \qquad \sum_{i,j=1}^{n} \fint_Q [(\nabla_j\phi_i)^2 - (\nabla_i\phi_i)(\nabla_j\phi_j)]\,dy,$$

since the terms which are linear in $\nabla\phi$ vanish. Integration by parts gives

$$\fint_Q [(\nabla_j\phi_i)(\nabla_i\phi_j) - (\nabla_i\phi_i)(\nabla_j\phi_j)]\,dy = 0$$

for each i and j. Therefore (5.6) equals

We shall apply duality to (5.11). If $B > \gamma I$ then Φ_B is convex, and can be written as a supremum of linear functions

$$(5.12) \qquad \Phi_B(\xi) = \sup_\eta \langle \xi, \eta \rangle - \Phi_B^*(\eta).$$

The Legendre transformation Φ_B^* is again quadratic, and it is characterized by

$$(5.13) \qquad \Phi_B^*(\eta) = \sup_\xi \langle \xi, \eta \rangle - \Phi_B(\xi).$$

We now restrict γ to the range $0 < \gamma < \alpha$, so that $a(y) > \gamma$ for each y. Then (5.11) and (5.12) yield

$$\Phi_A(\xi) \geq \inf_\phi \sup_\eta \int_Q \langle \xi + \nabla\phi, \eta \rangle - \Phi_{a(y)I}^*(\eta) \, dy,$$

in which η ranges over periodic, square-integrable, matrix-valued functions. Interchanging the inf and sup - appealing as usual to [7] or to a direct argument - and evaluating the infimum over ϕ, we get

$$(5.14) \qquad \Phi_A(\xi) \geq \sup_{\mathrm{div}\ \eta = 0} \int_Q [\langle \xi, \eta \rangle - \Phi_{a(y)I}^*(\eta)] \, dy,$$

where "div $\eta = 0$" means that the columns of η are divergence-free.

Restricting attention to <u>constant</u> test fields in (5.14), it follows that

$$\Phi_A(\xi) - \langle \xi, \eta \rangle \geq - \int_Q \Phi_{a(y)I}^*(\eta) \, dy$$

for any $n \times n$ matrices ξ and η. Minimizing over ξ on the left, evaluating the integral on the right, and rearranging, we finally conclude that

$$(5.15) \qquad \Phi_A^*(\eta) \leq \theta_\alpha \Phi_{\alpha I}^*(\eta) + \theta_\beta \Phi_{\beta I}^*(\eta).$$

This gives infinitely many bounds on A, one for each η. To make them explicit we must calculate $\Phi_A^*(\eta)$. The optimal ξ for (5.13) (with B replaced by A) satisfies

$$(5.16) \qquad (A-\gamma)\xi + \gamma(\text{tr }\xi)I = \eta \ ,$$

and the extremal value is half the linear term

$$(5.17) \qquad \Phi_A^*(\eta) = \tfrac{1}{2} \ \langle\xi,\eta\rangle \ .$$

The general inversion of (5.16) is messy, but the special case $\eta = I$ is easy: then (5.16) yields

$$\xi + \gamma(\text{tr}\xi)(A-\gamma I)^{-1} = (A - \gamma I)^{-1}$$

whence

$$\text{tr}\xi \cdot (1 + \gamma \ \text{tr}[(A-\gamma I)^{-1}]) = \text{tr} \ [(A-\gamma I)^{-1}],$$

and (5.17) leads to

$$(5.18) \qquad \Phi_A^* (I) = \tfrac{1}{2} \ \text{tr}\xi = \tfrac{1}{2} \ \frac{\text{tr}[(A-\gamma I)^{-1}]}{1+\gamma \text{tr}[(A-\gamma I)^{-1}]} \ .$$

We need not calculate $\Phi_A^* (\eta)$ for other choices of η, because $\eta = I$ gives the desired lower bound: substitution of (5.18) into (5.15) yields

$$\frac{\text{tr}[(A-\alpha I)^{-1}]}{1+\alpha \text{tr}[(A-\alpha I)^{-1}]} \leqslant \theta_\alpha \cdot \frac{1}{\alpha} + \theta_\beta \cdot \frac{n}{\beta-\alpha + \alpha n} \ ,$$

after passage to the limit $\gamma\!\uparrow\!\alpha$. This is equivalent to (LB) by straightforward algebraic manipulation.

5D. The upper bound.

The argument is parallel to that for the lower bound. Given a matrix $\overline{\sigma}$, let $\hat{\sigma}$ be the minimizer of (5.2). Then

$$\langle A^{-1}\overline{\sigma}, \overline{\sigma}\rangle = \int_Q a(y)^{-1}|\hat{\sigma}|^2 dy$$

$$\geqslant \int_Q [a(y)^{-1}|\hat{\sigma}|^2 - \gamma \ S(\hat{\sigma})] dy + \gamma \ S(\overline{\sigma})$$

for any $\gamma > 0$, by (5.4), whence

$$\langle A^{-1}\overline{\sigma}, \overline{\sigma}\rangle - \gamma S(\overline{\sigma}) \geq \inf_{\substack{\text{div } \sigma=0 \\ f_Q\ \sigma=\overline{\sigma}}} f_Q[a(y)^{-1}|\sigma|^2 - \gamma S(\sigma)]dy.$$

This can be rewritten as

$$(5.19) \qquad \Psi_A(\overline{\sigma}) \geq \inf_{\substack{\text{div}\sigma=0 \\ f_Q\ \sigma=\overline{\sigma}}} f_Q\ \Psi_{a(y)I}(\sigma)\ dy,$$

with

$$\Psi_B(n) = \frac{1}{2}\langle B^{-1}-\gamma)n,n\rangle + \frac{\gamma}{2(n-1)}\ (\text{tr}\,n)^2$$

for any $n \times n$ matrix n and any symmetric B.

We shall dualize (5.19). First rewrite it as

$$(5.20) \qquad \Psi_A(\overline{\sigma}) \geq \inf_\sigma \sup_{\xi,\phi} f_Q[\Psi_{a(y)I}(\sigma) - \langle\sigma-\overline{\sigma},\xi\rangle - \langle\sigma,\nabla\phi\rangle]dy,$$

in which σ ranges over periodic, matrix valued functions; ϕ over periodic, vector-valued functions; and ξ over constant matrices. Next write $\Psi_{a(y)I}(\sigma)$ as a supremum of linear functions: we have

$$(5.21) \qquad \Psi_B(\sigma) = \sup_n \langle n,\sigma\rangle - \Psi_B^*(n)$$

with

$$(5.22) \qquad \Psi_B^*(\xi) = \sup_\sigma \langle\xi,\sigma\rangle - \Psi_B(\sigma)$$

whenever $B^{-1} \geq \gamma I$. Restricting γ to the range $0 < \gamma < \beta^{-1}$, and using (5.21) with $B = a(y)I$, the right side of (5.20) becomes

$$\inf_\sigma \sup_{\xi,\phi,n} f_Q[\langle n - \xi - \nabla\phi,\sigma\rangle + \langle\overline{\sigma},\xi\rangle - \Psi_{a(y)I}^*(n)]dy\ .$$

Duality lets us switch the inf and the sup, and the infimum over σ is $-\infty$ unless $n = \xi + \nabla\phi$, so

$$(5.23) \qquad \Psi_A(\overline{\sigma}) \geq \sup_{\xi,\phi} f_Q[\langle\overline{\sigma},\xi\rangle - \Psi_{a(y)I}^*(\xi+\nabla\phi)]dy.$$

We take $\phi = 0$ in (5.23) to get

$$\Psi_A(\overline{\sigma}) - \langle \overline{\sigma}, \xi \rangle \geqslant - \int_0 \Psi^*_{a(y)I}(\xi)dy$$

for each pair of matrices $\overline{\sigma}$, ξ. Minimizing over $\overline{\sigma}$ on the left, evaluating the integral on the right, and rearranging, it follows finally that

$$(5.24) \qquad \Psi^*_A(\xi) \leqslant \theta_\alpha \Psi^*_{\alpha I}(\xi) + \theta_\beta \Psi^*_{\beta I}(\xi).$$

Like (5.15), this gives infinitely many bounds on A, one for each ξ; and as before, the one we want is obtained when $\xi = I$. Indeed, for this ξ the minimizer σ of (5.22) (with B replaced by A) satisfies

$$I = (A^{-1} - \gamma I)\sigma + \frac{\gamma}{n-1}(\text{tr}\sigma) I,$$

whence

$$(1 - \frac{\gamma}{n-1} \text{tr } \sigma) \cdot \text{tr } ([A^{-1} - \gamma I]^{-1}) = \text{tr } \sigma.$$

The extremal value of (5.22) is half the linear term,

$$(5.25) \qquad \Psi^*_A(I) = \frac{1}{2} \langle I, \sigma \rangle = \frac{1}{2} \frac{(n-1)\text{tr}[A(I-\gamma A)^{-1}]}{(n-1)+\gamma\text{tr}[A(I-\gamma A)^{-1}]}.$$

Substitution in (5.24) gives, after passage to the limit $\gamma \uparrow \beta^{-1}$,

$$\frac{(n-1)\beta\text{tr}[A(\beta I-A)^{-1}]}{(n-1)+\text{tr}[A(\beta I-A)^{-1}]} \leqslant \theta_\alpha \frac{(n-1)n\alpha\beta}{(n-1)\beta+\alpha} + \theta_\beta \frac{(n-1)n\beta}{n}$$

This is equivalent to (UB) by algebraic manipulation, making use of the identity

$$\text{tr}[A(\beta I-A)^{-1}] = \beta\text{tr}([\beta I-A]^{-1}) - n.$$

§6. Bounds via Representation Formulas

This method, first introduced by D. Bergman [3], has been developed and extended by many authors; a recent review is [24]. Our exposition draws primarily from [9] and [10]. The starting point is a pair of representation formulas for the effective conductivity A ,

$$(6.1a) \qquad A - \alpha I = \alpha(\beta-\alpha) \int_0^1 \frac{d\mu(z)}{\alpha+(\beta-\alpha)z},$$

$$(6.1b) \qquad \beta I - A = \beta(\beta-\alpha) \int_0^1 \frac{d\nu(z)}{\beta+(\alpha-\beta)z},$$

in terms of two nonnegative, symmetric matrix-valued measures μ and ν . Perturbation theory around the case $\alpha=\beta$ leads to constraints on the measures:

$$(6.2a) \qquad \int_0^1 d\mu = \theta_\beta I \ , \qquad\qquad tr \left(\int_0^1 z d\mu\right) = \theta_\beta\theta_\alpha \ ,$$

$$(6.2b) \qquad \int_0^1 d\nu = \theta_\alpha I \ , \qquad\qquad tr \left(\int_0^1 z d\nu\right) = \theta_\beta\theta_\alpha \ .$$

These, in turn, lead by algebraic manipulation to the desired bounds.

6A. Representation formulas.

We shall establish (6.1a) for any $\alpha, \beta > 0$ (without assuming that $\alpha < \beta$). There is no need to prove (6.1b) separately, since it follows from (6.1a) on interchanging the roles of α and β .

The starting point is the definition (3.4):

$$A\xi = \int_Q a(y)(\xi + \nabla\phi^\xi)dy,$$

where ϕ^ξ is the periodic solution of

$$div(a(y)(\xi + \nabla\phi^\xi)) = 0.$$

Since $a(y) = \alpha\chi_\alpha + \beta\chi_\beta$, the formula for A can be written

$$(6.3) \qquad (A-\alpha I)\xi = (\beta - \alpha)\int_Q \chi_\beta(\xi + \nabla\phi^\xi)dy.$$

The equation for ϕ^ξ becomes

$$(\beta - \alpha)\operatorname{div}(\chi_\beta(\xi + \nabla\phi^\xi)) = -\alpha\Delta\phi^\xi \ ,$$

or again

$$(6.4) \qquad (I + (\tfrac{\beta}{\alpha} - 1)\nabla\Delta^{-1}\operatorname{div}\chi_\beta)(\xi + \nabla\phi^\xi) = \xi \ .$$

A combination of (6.3) and (6.4) gives

$$(6.5) \qquad (A - \alpha I)\xi = (\beta - \alpha)f_Q\,\chi_\beta(I + (\tfrac{\beta}{\alpha} - 1)\Lambda)^{-1}\xi \ dy \ ,$$

where Λ is the linear operator

$$(6.6) \qquad \Lambda\eta = \nabla\Delta^{-1}\operatorname{div}(\chi_\beta\eta),$$

acting on the space L^2_{per} of Q-periodic, square-integrable vector fields. One verifies easily that Λ is self-adjoint with respect to the inner product

$$[\eta,\lambda] = f_Q\,\chi_\beta\,\langle\eta(y),\lambda(y)\rangle \ dy \ ,$$

and that

$$0 < [\eta, \Lambda\eta] < [\eta,\eta].$$

Therefore the spectrum of Λ lies in $[0,1]$. The spectral theorem provides a family of projection-valued measures $dP(z)$ on $[0,1]$ such that

$$[f(\Lambda)\xi, \eta] = \int_0^1 f(z)\,[dP(z)\xi,\eta]$$

for any $\xi, \eta \in L^2_{per}$ and any bounded, continuous function f on $[0,1]$. Restricting attention to constant ξ and η, and choosing $f(z) = (1 + (\tfrac{\beta}{\alpha} - 1)z)^{-1}$, it follows that

$$f_Q\,\chi_\beta\,\langle(I+(\tfrac{\beta}{\alpha} - 1)\Lambda)^{-1}\xi,\eta\rangle \ dy = \int_0^1 \frac{[dP(z)\xi,\eta]}{1+(\tfrac{\beta}{\alpha} - 1)z} \ .$$

In view of (6.5), this give the desired representation formula with the measure μ defined by

$$\langle d\mu(z)\xi,\eta\rangle = [dP(z)\xi,\eta], \quad \xi,\eta \quad R^n.$$

It is clear from the construction that μ takes its value in the space of non-negative, symmetric matrices, and that it depends on the geometry (through χ_β) but not on the microscopic conductivities α and β.

6B. Perturbation theory.

We need only verify (6.2a), since (6.2b) is obtained by switching the roles of α and β. The series

$$(1+\varepsilon z)^{-1} = 1-\varepsilon z + \varepsilon^2 z^2 + \ldots$$

converges uniformly for $z \quad [0,1]$ when $|\varepsilon| < 1$. With $\varepsilon = \frac{\beta}{\alpha} - 1$, substitution into (6.1a) yields

$$(6.7) \qquad \frac{1}{\alpha} A - I = (\frac{\beta}{\alpha} - 1) \int_0^1 d\mu - (\frac{\beta}{\alpha} - 1)^2 \int_0^1 z d\mu + \ldots$$

On the other hand, the series

$$(I + \varepsilon \Lambda)^{-1} = I - \varepsilon \Lambda + \varepsilon^2 \Lambda + \ldots$$

converges when $|\varepsilon| < 1$, and substitution into (6.5) with $\varepsilon = \frac{\beta}{\alpha} - 1$ gives

$$(6.8) \qquad \frac{1}{\alpha} A - I = (\frac{\beta}{\alpha} - 1) \int_0 \chi_\beta I dy - (\frac{\beta}{\alpha} - 1)^2 \int_0 \chi_\beta \Lambda \, dy + \ldots \, .$$

Equating the coefficients of $(\frac{\beta}{\alpha} - 1)$ in (6.7) and (6.8) gives the first part of (6.2a):

$$\int_0^1 d\mu = \int_0 \chi_\beta I dy = \theta_\beta I.$$

Equating the coefficients of $(\frac{\beta}{\alpha} - 1)^2$ gives, for any $\xi \in R^n$,

$$\int_0^1 z \langle d\mu(z)\xi,\xi\rangle = \int_0 \chi_\beta \langle \xi, \Lambda \xi\rangle dy$$

$$= \int_0 \langle \chi_\beta \xi, \, \nabla \Delta^{-1} \mathrm{div} \, \chi_\beta \xi\rangle dy \, ,$$

using the definition (6.6) of Λ. This last expression is just $\langle F\xi,\xi\rangle$, the quadratic form of geometric parameters (4.1), with $\chi = \chi_\beta$. Its trace is $\theta_\alpha \theta_\beta$, by Lemma 4.1, and this gives the second part of (6.2a).

6C. An inequality on positive measures.

Relations (6.2a,b) constrain the masses and first moments of μ and ν. The main tool for drawing conclusions about A is the following lemma concerning real-valued measures.

__Lemma 6.1.__ __Let__ m __be a positive measure on__ $[0,1]$ __with__

$$(6.9) \qquad \int_0^1 dm = m_0 \ , \quad \int_0^1 z\, dm = m_1.$$

__If__ $s>0$ __and__ $s + t > 0$ __then__

$$\int_0^1 \frac{dm(z)}{s+tz} \ \geqslant \ m_0 \left(s + \frac{t m_1}{m_0}\right)^{-1}.$$

__Proof:__ We may suppose $t \neq 0$, since the case $t = 0$ is trivial. Clearly

$$(6.10) \qquad \int_0^1 \frac{dm(z)}{s+tz} = \int_0^1 \left(\frac{1}{s+tz} + \lambda z\right) dm - \lambda m_1$$

for any λ, by (6.9). If

$$(6.11) \qquad \lambda = t^{-1} \left(\frac{s}{t} + \frac{m_1}{m_0}\right)^{-2}$$

then

$$\inf_{0<z<1} \frac{1}{s+tz} + \lambda z = t^{-1}\left(\frac{s}{t} + \frac{m_1}{m_0}\right)^{-1}\left(1 + \frac{m_1}{m_0}\left(\frac{s}{t} + \frac{m_1}{m_0}\right)^{-1}\right),$$

achieved exactly when $z = m_1/m_0$. Since m has mass m_0,

$$\int_0^1 \frac{dm}{s+tz} \geqslant m_0 \cdot \inf_{0<z<1} \left(\frac{1}{s+tz} + \lambda z\right) - \lambda m_1$$

$$= m_0 t^{-1} \left(\frac{s}{t} + \frac{m_1}{m_0}\right)^{-1}.$$

The choice of λ in the preceding proof can be motivated by convex duality. Indeed, we seek

$$\inf_{\int_0^1 dm \,=\, m_0} \left\{ \sup_\lambda \; \int_0^1 \frac{dm}{s+tz} \;+\; \lambda\left(\int_0^1 z\,dm - m_1\right) \right\},$$

where λ serves as a Lagrange multiplier for the constraint $\int_0^1 z\,dm = m_1$. Interchanging the inf and sup leads to the dual problem

$$\sup_\lambda \; \left\{ \inf_{\int_0^1 dm \,=\, m_0} \; \int_0^1 \left(\frac{1}{s+tz} + \lambda z\right)dm \right\} - \lambda m_1.$$

Calculating the infimum explicitly then maximizing the result over λ leads to (6.11).

6D. The bounds.

We now establish the lower bound (LB), as a consquence of (6.1a) and (6.2a). Fixing α and β, let us work in a basis of eigenvectors of A, denoting by A_i the ith eigenvalue of A. We shall write μ_i instead of μ_{ii} for the ith diagonal element of μ. By (6.1a),

$$A_i - \alpha = \alpha(\beta-\alpha) \int_0^1 \frac{d\mu_i}{\alpha+(\beta-\alpha)z} \qquad ;$$

also, by (6.2a),

$$\int_0^1 d\mu_i = \theta_\beta \;,\; \int_0^1 z\,d\mu_i = F_i$$

with

$$(6.12) \qquad \sum_{i=1}^n F_i = \theta_\alpha \theta_\beta$$

Lemma 6.1 gives

$$A_i - \alpha \,>\, \alpha\theta_\beta \left(\frac{\alpha}{\beta-\alpha} + \frac{F_i}{\theta_\beta}\right)^{-1}$$

whence

$$(A_i - \alpha)^{-1} \leq \alpha^{-1}\theta_\beta^{-1}\left(\frac{\alpha}{\beta-\alpha} + \frac{F_i}{\theta_\beta}\right).$$

Summation over i gives

$$(6.13) \qquad \text{tr}[(A-\alpha I)^{-1}] \leq \frac{n}{\theta_\beta(\beta-\alpha)} + \frac{\theta_\alpha}{\alpha\theta_\beta},$$

which is identical to (4.6) and equivalent to (LB).

The upper bound is obtained similarly as a consequence of (6.1b) and (6.2b):

$$\beta - A_i = \beta(\beta - \alpha)\int_0^1 \frac{d\nu_i}{\beta+(\alpha-\beta)z}$$

with

$$\int_0^1 d\nu_i = \theta_\alpha \, , \quad \int_0^1 z d\nu_i = G_i \, , \sum_{i=1}^n G_i = \theta_\alpha\theta_\beta \, .$$

Lemma 6.1 gives

$$(\beta-A_i) \geq \beta\theta_\alpha\left(\frac{\beta}{\beta-\alpha} - \frac{G_i}{\theta_\alpha}\right)^{-1},$$

whence

$$(\beta-A_i)^{-1} \leq \beta^{-1}\theta_\alpha^{-1}\left(\frac{\beta}{\beta-\alpha} - \frac{G_i}{\alpha}\right)$$

and so

$$(6.14) \qquad \text{tr}[(\beta I-A)^{-1}] \leq \frac{n}{\theta_\alpha(\beta-\alpha)} - \frac{\theta_\beta}{\beta\theta_\alpha},$$

the same as (4.8) and equivalent to (UB).

This proof explains the symmetry between the right hand sides of (LB) and (UB). Each becomes minus the other when the roles of α and β are exchanged (c.f. (6.13) and (6.14)). It must be so, because they follow from (6.1a) - (6.2a) and (6.1b) - (6.2b), which are related similarly.

§7. Comparison of the Methods.

We have presented three different methods for bounding the effective conductivity of a (generally anisotropic) composite. The first combines the Hashin-Shtrikman variational principles with linear relations on a tensor F of geometric parameters; the second uses lower semicontinuous, nonconvex quadratic forms; and the third combines representation formulas with perturbation theory. In the case discussed here - mixtures of two isotropic conductors in R^n - all three approaches, different as they are, lead to the same (optimal) result. In their generalizations, however, the methods diverge.

The first one, using the Hashin-Shtrikman variational principles, extends directly to a large class of problems; [17] discusses its use for multicomponent composites with anisotropic components, in the context of both conductivity and linear elasticity. Except for the choice of a reference medium there is very little flexibility - which makes the method easy to execute (an advantage), but restricts the range of conclusions (possibly a disadvantage). When both the reference medium and the composite itself are isotropic, this approach <u>coincides</u> with that executed by Hashin and Shtrikman in the 1960's.

The second method, using compensated compactness, is also easily extended to multicomponent composites with anisotropic components, and to systems such as linear elasticity. Being new, it has been applied as yet to just a few situations. In linear elasticity, it has given a generalization of the Hashin-Shtrikman bound on the bulk modulus of an isotropic mixture of isotropic materials [27]. It has also been used successfully in plate theory [8]. In general, it is less automatic than the variational approach: the class of lower semicontinuous, quadratic functionals is larger than the class of reference media. This presents a problem of choice: the method is flexible (an advantage) but allows a plethora of possibilities (a disadvantage). Perhaps more experience will provide better guidance on its use.

The situation is quite different as concerns our third method, using representation formulas. Unlike the other two, it makes essential use of the hypothesis that there are only two components. However, again unlike the others, it yields information on <u>complex</u> conductivities, by analytic continuation of the representation formulas into the complex plane. This has been thoroughly explored; see [9,24,39] and the references cited there. A similar method has been applied to linear elasticity in [15], and a more sophisticated representation formula has been derived for use in the multicomponent case [11]. These remain areas of current research activity.

Since the three methods lead in many cases to the same bounds, it is natural to look for direct relations between them. An equivalence can indeed be drawn between the variational and representation formula methods [23], at least for two-component composites in the context of conductivity. A similar connection to the compensated compactness method has yet to be found.

Nonlinear problems present an important, largely uncharted territory for future exploration. If each component material is characterized by a convex variational principle, then the composite will behave so as to minimize a convex <u>effective</u> <u>energy</u>. Physical applications include both nonlinear conductors and Hencky plasticity. The Hashin-Shtrikman variational method is applied to such a problem in [31,37], yielding upper and lower bounds for the effective energy. It would be interesting to know what the compensated compactness approach gives for the same problem.

<u>Acknowledgements</u>: This work was begun while the authors were visiting the Institute for Mathematics and its Applications. RVK gratefully acknowledges additional support from NSF grant DMS-8312229, ONR grant N00014-83-0536, and the Sloan Foundation. GM gratefully acknowledges support from Caltech, through a Weingart Fellowship, and from Chevron Laboratories, through a research grant.

References

1. Armand, J.-L. Lurie, K.A., and Cherkaev, A.V., "Optimal control theory and structural design," in Optimum Structure Design Vol. 2, R.H. Gallagher; E. Atrek, K. Ragsdell, and O.C. Zienkiewicz, eds., J. Wiley and Sons, 1983.

2. Bensoussan, A., Lions. J.-L., and Papanicolaou, G., Asymptotic Analysis for Periodic Structures, North-Holland, 1978.

3. Bergman, D., "The dielectric constant of a composite material - a problem in classical physics" Phys. Rep. C 43, 1978, pp. 377-407.

4. Christensen, R.M., Mechanics of Composite Materials, Wiley Interscience, 1979.

5. Dacorogna, B., Weak Continuity and Weak Lower Semicontinuity of Nonlinear Functionals, Lecture Notes in Math. 922, Springer-Verlag, 1982.

6. De Giorgi, E. and Spagnolo, S., "Sulla convergenza delli integrali dell'energia per operatori ellittici del secondo ordine," Boll. Un. Mat. Ital. 8, 1978, pp. 291-411.

7. Ekeland, I. and Temam, R., Convex Analysis and Variational Problems, North-Holland, 1976.

8. Gibiansky, L.V. and Cherkaev, A.V., "Design of composite plates of extremal rigidity," preprint.

9. Golden, K., "Bounds on the complex permittivity of a multicomponent material" J. Mech. Phys. Solids, to appear.

10. Golden, K. and Papanicolaou, G., "Bounds for effective parameters of heterogeneous media by analytic continuation," Comm. Math. Phys. 90, 1983, pg. 473.

11. Golden, K., and Papanicolaou, G., to appear in J. Stat. Phys..

12. Hashin, Z. "Analysis of composite materials - a survey," J. Appl. Mech. 50, 1983, pp. 481-505.

13. Hashin, Z. and Shtrikman, S., "A variational approach to the theory of the effective magnetic permeability of multiphase materials," J. Appl. Phys. 33, 1962, pp. 3125-3131.

14. Hill, R., "New derivations of some elastic extremum principles," in Progress in Applied Mechanics - Prager Aniversary Volume, Mac Millan, 1963, pp. 99-106.

15. Y. Kantor and D. Bergman, "Improved rigorous bounds on the effective elastic moduli of a composite material" J. Mech. Phys. Solids 32, 1984, pp. 41-62.

16. Kohn, R. and Dal Maso, G., in preparation.

17. Kohn, R.V. and Milton, G.W., "Bounds for anisotropic composites by variational principles," in preparation.

18. Kohn, R.V. and Strang, G., "Structural design optimization, homogenization, and relaxation of variational problems" in Macroscopic Properties of Disordered Media, R. Burridge, S. Childress, G. Papanicolaou eds., Lecture Notes in Physics 154, Springer-Verlag, 1982, pp. 131-147.

19. Kohn, R.V. and Strang, G., "Optimal design and relaxation of variatonal problems," to appear in Comm. Pure Appl. Math..

20. Lurie, K.A. and Cherkaev, A.V., "Optimal structural design and relaxed controls", Opt. Control Appl. and Meth. 4, 1983, pp. 387-392.

21. Lurie, K.A. and Cherkaev, A.V., "Exact estimates of conductivity of composites formed by two isotropically conducting media taken in prescribed proportion," Proc. Royal Soc. Edinburgh 99A, 1984, pp. 71-87.

22. Lurie, K.A. and Cherkaev, A.V., "Exact estimates of the conductivity of a binary mixture of isotropic compounds," preprint.

23. Milton, G.W. and McPhedran, R.C., "A comparison of two methods for deriving bounds on the effective conductivity of composites," in Macroscopic Properties of Disordered Media, R. Burridge, S. Childress, G. Papanicolaou, eds., Lecture Notes in Physics 154, Springer-Verlag, 1982, pp. 183-193.

24. Milton, G.W. and Golden, K., "Thermal conduction in composites" in Thermal Conductivity 18, Plenum Press, 1985.

25. Murat, F., H-convergence, mimeographed notes, 1978.

26. Murat, F., "Compacité par compensation," Ann. Scuola Norm. Sup. Pisa 5, 1978, pp. 489-507.

27. Murat, F. and Francfort, G.A., "Homogenization and optimal bounds in linear elasticity," to appear in Arch. Rat. Mech. Anal.

28. Murat, F. and Tartar, L., "Calcul des variations et homogénéisation", in Les Methodes de l'Homogénéisation: Theorie et Applications en Physique, Coll. de la Dir. des Etudes et Recherches de Electricité de France, Eyrolles, Paris, 1985, pp. 319-370.

29. Papanicolaou, G. and Varadhan, S., "Boundary value problems with rapidly oscillting random coefficients," in Colloquia Mathematica Societatis János Bolyai 27, Random Fields, North-Holland, 1982, pg. 835.

30. Sanchez-Palencia, E., Non-homogeneous Media and Vibration Theory, Lecture Notes in Physics 127, Springer-Verlag, 1980.

31. Talbot, D.R.S. and Willis, J.R., "Variational Principles for inhomogeneous nonlinear media," to appear in IMA J. Appl. Math. .

32. Tartar, L., "Compensated compactness and applications to P.D.E.," in Nonlinear Analysis and Mechanics: Heriot-Watt Symposium IV, R. Knops, ed., Pitman Press, 1979, pp. 136-212.

33. Tartar, L., "Estimations fines des coefficients homogénéises," in Ennio De Giorgi Colloquium, P. Krée, ed., Pitman Press, 1985. See also the article by Tartar in this volume.

34. Willis, J.R., "Bounds and self-consistent estimates for the overall moduli of anisotropic composites," J. Mech. Phys. Solids 25, 1977, pp. 185-202.

35. Willis, J.R., "Variational and related methods for the overall properties of composite materials," in C.-S. Yih, ed., Advances in Applied Mechanics 21, 1981, pp. 2-78.

36. Willis, J.R., "The overall elastic response of composite materials," J. Appl. Mech. 50, 1983, pp. 1202-1209.

37. Willis, J.R, "Variational estimates for the overall response of an inhomo-geneous nonlinear dielectric," this volume.

38. Zhikov, V.V., Kozlov, S.M., Oleinik, O.A., and Ngoan, K.T., "Averaging and G-convergence of differential operators" _Russian Math. Surveys_, 34, 1979, pp. 69-147.

39. Bergman, D., "Rigorous bounds for the complex dielectric constant of a two-component composite" _Annals of Physics_ 138, 1982, pg. 78.

THIN PLATES WITH RAPIDLY VARYING THICKNESS, AND THEIR RELATION TO STRUCTURAL OPTIMIZATION

By

Robert V. Kohn
Courant Institute of Mathematical Sciences
New York, NY 10012

and

Michael Vogelius
Department of Mathematics and
Institute for Physical Science and Technology
University of Maryland
College Park, MD 20742

Introduction

There is a close relationship between problems of structural optimization
and the analysis of media with microstructure. The optimal design of variable
thickness plates is a case in point: for certain problems, plates with
"stiffeners" formed by rapid thickness variation can be stronger per unit volume
than any traditional, uniform or slowly varying plates. To resolve such a
design problem one must introduce a "generalized plate model," representing the
overall effect of a microstructure of stiffeners on the behavior of the plate.

One idea would be to substitute a rapidly varying thickness function into
the fourth-order equation of Kirchhoff plate theory and perform some kind of
"homogenization". There is, however, a physically more correct approach: it
appeals directly to three-dimensional linear elastostatics on thin, rapidly-varying,
plate-like domains. There are two small parameters -- the mean thickness ε
and the length scale of thickness variation δ -- and one can study the asymptotics of the solution as they both tend to zero. This was the focus of our
recent papers [13, 14]. We showed that <u>it makes a difference which parameter
tends to zero faster</u>. Use of the Kirchhoff plate equation with a rapidly varying
thickness corresponds to the case $\varepsilon \ll \delta$. The other extreme, $\delta \ll \varepsilon$,
corresponds to averaging the effect of the thickness variation first, then
applying Kirchhoff theory to the resulting anisotropic plate. Intermediate between

these is a third case, $\varepsilon \sim \delta$, which has no such simple interpretation. For applications to optimal design it is natural to ask which alternative gives the strongest structure, and that was the focus of our most recent paper [15].

The present article is an expository review of this work and its relevance to optimization. Special attention is focused on plates with "one family of stiffeners," for which the theory is relatively complete. Much remains to be done for more general thickness variation; various open questions will be indicated as we proceed, and especially in section 6. We shall refer only to the most recent relevant articles, without any attempt at a complete survey of the extensive literature. More references on homogenization and plate theory can be found in [6,13,24] and an extensive bibliography on structural optimization is given in [3]. Recent surveys on plate optimization include [2] and [20].

1. An Optimal Design Problem

Kirchhoff plate theory models the behavior of symmetric, variable-thickness plates under transverse loads. It specifies the vertical displacement w_0 as the solution of an elliptic equation

$$(1.1) \qquad \frac{\partial^2}{\partial x_\alpha \partial x_\beta} \left(M_{\alpha\beta\gamma\delta} \frac{\partial^2 w_0}{\partial x_\gamma \partial x_\delta} \right) = F$$

on the midplate domain ω, with appropriate boundary conditions at the plate edges $\partial\omega$. The tensor $M_{\alpha\beta\gamma\delta}$ relates bending moment to midplane curvature; it depends on the plate's thickness $2h$ and on the constant elastic moduli $B_{ijk\ell}$ of the material from which the plate is made, through the formula

$$(1.2) \qquad M_{\alpha\beta\gamma\delta} = \frac{2}{3} h^3 \, \widetilde{B}_{\alpha\beta\gamma\delta} \, ,$$

where

$$\widetilde{B}_{\alpha\beta\gamma\delta} = B_{\alpha\beta\gamma\delta} - B_{\alpha\beta 33} B_{\gamma\delta 33} \, / \, B_{3333} \, .$$

(The Hooke's law tensor $B_{ijk\ell}$ is assumed to satisfy the usual symmetries

$B_{ijk\ell} = B_{jik\ell} = B_{k\ell ij}$, and to have the midplane as a plane of elastic symmetry.) For an isotropic material, $\widetilde{B}$ is given by

$$\widetilde{B}_{1111} = \widetilde{B}_{2222} = E/(1-\nu^2)$$

$$\widetilde{B}_{1122} = \widetilde{B}_{2211} = E\nu/(1-\nu^2)$$

$$\widetilde{B}_{1212} = \widetilde{B}_{1221} = \widetilde{B}_{2112} = \widetilde{B}_{2121} = E/2(1+\nu),$$

where ν denotes Poisson's ratio and E is Young's modulus. The right side of (1.1) is the load per unit midplane area.

For simplicity, we shall discuss only plates that are clamped at the edges; this means that

$$w_0 = \frac{\partial w_0}{\partial n} = 0 \quad \text{at} \quad \partial\omega .$$

The principle of minimum energy gives an alternate characterization of w_0 as **the** minimizer of

$$(1.3) \qquad \frac{1}{2} \int_\omega M_{\alpha\beta\gamma\delta} \frac{\partial^2 w}{\partial x_\alpha \partial x_\beta} \frac{\partial^2 w}{\partial x_\gamma \partial x_\delta} - \int_\omega Fw$$

in the Sobolev space $\overset{\circ}{H}{}^2(\omega)$. The <u>compliance</u> L is the work done by the load,

$$L = \int_\omega Fw_0 = \int_\omega M_{\alpha\beta\gamma\delta} \frac{\partial^2 w_0}{\partial x_\alpha \partial x_\beta} \frac{\partial^2 w_0}{\partial x_\gamma \partial x_\delta} .$$

By (1.3), it has the variational characterization

$$L = - \int_\omega M_{\alpha\beta\gamma\delta} \frac{\partial^2 w_0}{\partial x_\alpha \partial x_\beta} \frac{\partial^2 w_0}{\partial x_\gamma \partial x_\delta} + 2 \int_\omega Fw_0$$

$$(1.4)$$

$$= \max_{w \in \overset{\circ}{H}{}^2(\omega)} \left(- \int_\omega M_{\alpha\beta\gamma\delta} \frac{\partial^2 w}{\partial x_\alpha \partial x_\beta} \frac{\partial^2 w}{\partial x_\gamma \partial x_\delta} + 2 \int_\omega Fw \right) .$$

For a given load F, we think of $L = L(h)$ as a functional of the (half)

thickness h. It represents an overall measure of the plate's rigidity under F.
Therefore it is natural to consider the problem of _optimization for minimum_
compliance: we seek to minimize L(h) among all plates with prescribed volume and
specified minimum and maximum thickness, _i.e._ among all h such that

$$(1.5) \qquad h \in L^{\infty}(\omega) \ , \ h_{min} \leqslant h \leqslant h_{max} \quad \text{and} \quad \int_{\omega} h dx = c.$$

It is now widely recognized that for some choices of F and h_{max}/h_{min} _this_
optimal design problem will have no solution. The difficulty is easy to
understand physically. We anticipate that formation of "stiffeners" by means of
an oscillatory thickness could improve the strength of the plate. Since tall,
thin beams are stronger than short, fat ones, the strength should increase as the
stiffener width tends to zero. If there is no optimum scale for the oscillation,
then there will be no optimal h. (A more precise version of the argument will be
presented in section 2.)

Numerical manifestations of this phenomenon have been observed in [1,9].
For certain loads F and sufficiently large ratios h_{max}/h_{min}, numerical methods
for minimizing L(h) are seen to display instabilities. The computed solutions
become strongly mesh-dependent, with "stiffeners" (oscillations of the thickness
between h_{min} and h_{max}) forming on the same scale as the mesh size.

Mathematically, the point is that L(h) is not weak* lower semicontinuous on
the space (1.5) of admissible h's. There will surely be a minimizing sequence $\{h_n\}$
which approaches the optimal behavior, and (after passage to a subsequence) it
will have a weak* limit h_∞. But the compliance can jump up in the limit, and in
that case h_∞ will not be an optimum.

Clearly there is something unsatisfactory about the formulation of a design
problem that has no solution. One way out is to _restrict_ the design space by
imposing a pointwise or integral bound on $|\nabla h|$ (cf. [5]). The other, we think
more natural alternative is to _extend_ the design space by allowing plates with
stiffeners or rapidly varying thickness [4,10,17]. This entails introduction of a
class $\mathcal{D}$ of "generalized plate-thicknesses" and an extension $\tilde{L}$ of L to $\mathcal{D}$
such that

(1.6a)   For each $\tilde{h} \in \mathcal{D}$ the generalized compliance $\tilde{L}(\tilde{h})$ is reali-
zable by a limit of ordinary plates. In other words,
there exists a sequence $\{h_n\}$ satisfying (1.5) for which
$\tilde{L}(\tilde{h}) = \lim_{n \to \infty} L(h_n)$.

(1.6b)   The functional $\tilde{L}$ attains its minimum value on $\mathcal{D}$.

The first condition assures that the stiffeners have been modelled correctly,
and hence that the underlying problem has not been altered. In particular, it
implies that $\inf L = \inf \tilde{L}$. The second condition says that the class $\mathcal{D}$ of
generalized thickness variations is "large enough". It promises that nothing
would be gained (for this design problem) by considering further extensions of the
design space.

The new problem of minimizing L on $\mathcal{D}$ is sometimes called a <u>full</u> <u>relaxa-</u>
<u>tion</u> of the original design problem. (The reader is warned, however, that this
term is used slightly differently in the calculus of variations, for example in
[12].) An extension to some intermediate class of plate models satisfying (1.6a)
but not (1.6b) could be called a <u>partial</u> <u>relaxation.</u> Finding a partial relaxation
requires the correct modelling of a particular class of plates with rapidly
varying thickness. Finding a full relaxation is more difficult: it requires
understanding just which types of stiffeners or rapidly varying thicknesses can
occur in an optimal structure. This remains in general an unsolved problem, but
the easier case of plates with a "single family of stiffeners" is fairly well in
hand. We shall discuss it in the next section.

As if finding a relaxation of the original design problem were not trouble
enough, there is also the further difficulty of its relation to three-dimensional
elasticity. This will be treated in sections 3 and 4, where we describe a class
of three-dimensional "plates" with rapidly-varying thickness which are correctly
modelled by homogenization of the Kirchhoff plate equation (1.1). The analysis
shows, however, that use of the Kirchhoff theory above represents a loss of information: plates with more rapid thickness variation require a different model.

Section 5 discusses the implications of this for structural optimization.

Though our discussion of the need for relaxation has focused on questions of existence, the relaxed problem is as important for computation as it is for the theory. Even partial relaxation may be advantageous for numerical use. Numerical minimization of a fully relaxed $\hat{L}$ will be free of the instabilities experienced using L; also, experience suggests that $\hat{L}$ will have fewer local minima than L. Finally, since $\hat{L}$ is known to achieve its minimum, one can obtain qualitative information about extremal designs by studying the first-order optimality conditions for $\hat{L}$.

2. Rapid Variation and Relaxation of the Compliance Functional

In order to relax the design problem, we must consider how rapid variations in h affect the compliance. There is a general theory of homogenization of periodic structures, which addresses precisely this sort of question [6,24]. It characterizes the vertical midplane displacement $\tilde{w}_0$ - in the limit as the length scale of the oscillation tends to zero - as the solution of (1.1) with a new, effective rigidity $\underline{\tilde{M}_{\alpha\beta\gamma\delta}}$. The limiting compliance is correspondingly $\int_\omega F\tilde{w}_0$.

The simplest case is that of a plate made from an isotropic material using "stiffeners in the x_2 direction." This means that h is a function of x_1 only, independent of x_2. We obtain oscillations on a length scale δ by taking the particular form

$$(2.1) \qquad h_\delta = H(x_1, x_1/\delta)$$

where $H(x_1, n_1)$ is periodic in the second variable with period 1, and sufficiently smooth in the first variable. If w_δ is the solution of (1.1) with $h = h_\delta$, then it is an exercise in homogenization to see that w_δ tends, as $\delta \to 0$, to the solution $\tilde{w}_0$ of

$$(2.2) \qquad \frac{\partial^2}{\partial x_\alpha \partial x_\beta} \left(\tilde{M}_{\alpha\beta\gamma\delta} \frac{\partial^2 \tilde{w}_0}{\partial x_\gamma \partial x_\delta} \right) = F$$

with

$$\tilde{M}_{1111} = \frac{2}{3} \frac{E}{1-\nu^2} \overline{H(x_1,\cdot)^{-3}}^{-1}$$

$$\tilde{M}_{2222} = \frac{2}{3} E \overline{H(x_1,\cdot)^3} + \frac{2}{3} \frac{E\nu^2}{1-\nu^2} \overline{H(x_1,\cdot)^{-3}}^{-1}$$

(2.3)

$$\tilde{M}_{1122} = \tilde{M}_{2211} = \frac{2}{3} \frac{E\nu}{1-\nu^2} \overline{H(x_1,\cdot)^{-3}}^{-1}$$

$$\tilde{M}_{1212} = \tilde{M}_{2112} = \tilde{M}_{1221} = \tilde{M}_{2121} = \frac{E}{3(1+\nu)} \overline{H(x_1,\cdot)^3} \ .$$

Here $\overline{H(x_1,\cdot)^3}$ denotes the average of the periodic function $H(x_1,\cdot)^3$ with respect to its second variable, and similarly for $\overline{H(x_1,\cdot)^{-3}}$. If $H = H(x_1)$ is independent of n_1, _i.e._ if there is no rapid variation, then (2.3) naturally agrees with (1.2). The convergence of w_δ towards $\tilde{w}_0$ is in the weak topology on $\overset{o}{H}{}^2$; it follows that the compliances converge

(2.4)
$$L(h_\delta) = \int_\omega F \, w_\delta \ \rightarrow \ \int_\omega F \, \tilde{w}_0 = \tilde{L} \ ,$$

and also that $w_\delta \rightarrow \tilde{w}_0$ uniformly on ω .

To see the advantage of rapid thickness variation, we consider oscillatory perturbations of a smoothly varying $\overline{H}$:

$$H_\lambda(x_1, n_1) = \overline{H}(x_1) + \lambda\phi(n_1) \ , \quad \phi \text{ is bounded and } \int_0^1 \phi(n_1)dn_1 = 0$$

If $h_{min} < \overline{H} < h_{max}$ then the same will be true for H_λ when λ is small enough. Calculation gives that

$$\frac{d}{d\lambda} \overline{H_\lambda^3} = 0, \quad \frac{d^2}{d\lambda^2} \overline{H_\lambda^3} = 6 \overline{H} \, \overline{\phi^2}$$

$$\frac{d}{d\lambda} (\overline{H_\lambda^{-3}})^{-1} = 0 \ , \quad \frac{d^2}{d\lambda^2} (\overline{H_\lambda^{-3}})^{-1} = -12 \overline{H} \, \overline{\phi^2}$$

at $\lambda = 0$. We see from (2.3) that a small, oscillatory perturbation always decreases $\tilde{M}_{1111}$ and $\tilde{M}_{1122}$, while it increases $\tilde{M}_{1212}$. In the physical range $0 < \nu < \frac{1}{2}$, $0 < E$, it also increases $\tilde{M}_{2222}$. This is entirely reasonable, since stiffeners should resist twisting and lengthwise bending, but should be rather weak under bending in the orthogonal direction.

We assert that for suitable loads F, the compliance is decreased by such perturbations. Indeed, let $\widetilde{M}_{\alpha\beta\gamma\delta}$ be the effective rigidity of the (oscillatory) perturbed geometry, $M_{\alpha\beta\gamma\delta}$ that of the (smooth) $\overline{H}$, and suppose that the solution $\widetilde{w}_\lambda$ of (2.2) satisfies

$$\left|\frac{\partial^2\widetilde{w}_\lambda}{\partial x_1^2}\right|^2 \ll \left|\frac{\partial^2\widetilde{w}_\lambda}{\partial x_2^2}\right|^2 + \left|\frac{\partial^2\widetilde{w}_\lambda}{\partial x_1 \partial x_2}\right|^2 .$$

Then the perturbed compliance $\widetilde{L}$ satisfies

$$\widetilde{L} = -\int_\omega \widetilde{M}_{\alpha\beta\gamma\delta} \frac{\partial^2\widetilde{w}_\lambda}{\partial x_\alpha \partial x_\beta} \frac{\partial^2\widetilde{w}_\lambda}{\partial x_\gamma \partial x_\delta} + 2 \int_\omega F\widetilde{w}_\lambda$$

$$< -\int_\omega M_{\alpha\beta\gamma\delta} \frac{\partial^2\widetilde{w}_\lambda}{\partial w_\alpha \partial x_\beta} \frac{\partial^2\widetilde{w}_\lambda}{\partial x_\gamma \partial x_\delta} + 2 \int_\omega F\widetilde{w}_\lambda$$

$$< L(\overline{H}) ,$$

and therefore the oscillations have improved the compliance. Intuitively, deformations of the desired type are expected (at least away from the edge of the plate) whenever F oscillates rapidly enough with respect to x_2.

The preceeding discussion is easily extended to a "single family of stiffeners" with smoothly varying profile and direction. This corresponds to choosing

$$h_\delta = H(\underset{\sim}{x}; \; e(\underset{\sim}{x})\cdot\underset{\sim}{x}/\delta) \;\; , \;\; |e(\underset{\sim}{x})|^2 = 1,$$

in place of (2.1). The stiffeners are then orthogonal to the field of unit vectors $e(\underset{\sim}{x})$; their profile is determined by $H(x_1,x_2;n)$, which should be periodic in the real variable n with period 1. Since homogenization is local, the effective rigidity at any $\underset{\sim}{x} \in \omega$ will be given by (2.3) in the orthogonal coordinate system which takes the x_1 axis parallel to $e(\underset{\sim}{x})$. The case of axially symmetric plates with circumferential stiffeners is an especially natural one. It was treated in [9,10], and rapid thickness variation was found to be advantageous for loads of the form $F = \cos k\theta$ when $k \geq 4$.

Returning for simplicity to plates with stiffeners in the x_2 direction (i.e. h a function of x_1 alone), we formulate a relaxation by following the ideas of [10]. The space $\mathcal{D}$ of "generalized thickness variations" consists of pairs $\tilde{h} = (q,\theta)$. The first element is restricted by $h_{min} \leq q(x_1) \leq h_{max}$; it represents the "minimum height of the stiffeners." The second is constrained by $0 \leq \theta(x_1) \leq 1$, and plays the role of the stiffener density. Both q and θ are assumed to be measurable, but not necessarily continuous. The rigidity tensor $\tilde{M}_{\alpha\beta\gamma\delta}$ corresponding to $\tilde{h}$ is defined by (2.3) with

$$
H(x_1,n_1) = \begin{cases} h_{max} & 0 < n_1 \leq \theta(x_1) \\ q(x_1) & \theta(x_1) < n_1 \leq 1 \, , \end{cases}
$$

and the corresponding compliance $\tilde{L}(\tilde{h})$ is obtained by solving (2.2).

This defines at least a partial relaxation, by virture of (2.4). We believe that it is the full relaxation, though this has not yet been established with mathematical rigor. Optimality conditions for the analogous relaxation of axi-symmetric plates are presented in [10]; they determine the circumstances under which "stiffeners" occur in an optimal structure.

The situation becomes much more complicated if more general thickness variation is allowed. For "two or more families of stiffeners" the effective rigidity can only be expressed in terms of the solutions of certain fourth-order equations on the period cell (see e.g. [13]). For a "single family of stiffeners" with unspecified direction, there is the possibility of the direction itself becoming oscillatory. And there are more general geometries to consider, analogous to the composites of rank two or more considered in [16], see also [23]. One route to a full relaxation involves seeking "optimal bounds" for the effective rigidity of a plate in terms of h_{min}, h_{max}, and the mean thickness $\bar{h}$; another involves the study of lower semicontinuity and the calculus of variations. Both approaches have successfully treated related problems [12,17,19,21] and it seems reasonable to hope for further progress soon.

3. An Apparent Physical Contradiction.

The primary justification for the plate equation (1.1) is that it follows from the equations of linear elasticity on the three-dimensional plate domain in the limit as the thickness tends to zero. Specifically, let Ω_ϵ denote the plate domain

$$\Omega_\epsilon = \{(x_1,x_2,x_3): (x_1,x_2) \in \omega, \ |x_3| < \epsilon \, h(x_1,x_2)\} \ ,$$

and let $\underline{u}^\epsilon$ be the linearly elastic displacement, satisfying:

$$\operatorname{div} \sigma(\underline{u}^\epsilon) = 0 \quad \text{in} \quad \Omega_\epsilon$$

$$\sigma(\underline{u}^\epsilon) \cdot \underline{n}^\epsilon = \frac{1}{2} \epsilon^3 (0,0,F|n_3^\epsilon|) \quad \text{for} \quad |x_3| = \pm \ \epsilon h$$

$$\underline{u}^\epsilon = 0 \quad \text{for} \quad (x_1,x_2) \in \partial\omega$$

Here $\sigma(\underline{u}^\epsilon)$ denotes the stress $B_{ijk\ell} \, e_{k\ell}$ associated to the strain $e_{k\ell} = \frac{1}{2} (\partial u_k^\epsilon/\partial x_\ell + \partial u_\ell^\epsilon/\partial x_k)$, and $\underline{n}^\epsilon$ is the outward unit normal to Ω_ϵ . Using either the dual variational principles [18] or a direct asymptotic expansion of $\underline{u}^\epsilon$ [11], one can show that the (rescaled) energy converges as $\epsilon \to 0$ to that given by the Kirchhoff model:

$$\lim_{\epsilon \to 0} \epsilon^{-3}(\frac{1}{2} \int_{\Omega_\epsilon} \sigma(\underline{u}^\epsilon) \cdot e(\underline{u}^\epsilon) - \frac{1}{2} \epsilon^3 \int_{x_3 = \pm \epsilon h} F|n_3^\epsilon|u_3^\epsilon)$$

$$(3.2) \qquad = \frac{1}{2} \int_\omega M_{\alpha\beta\gamma\delta} \frac{\partial^2 w_0}{\partial x_\alpha \partial x_\beta} \frac{\partial^2 w_0}{\partial x_\gamma \partial x_\delta} - \int_\omega F \, w_0 \ ,$$

where w_0 solves (1.1).

The convergence (3.2) holds provided that h and F are fixed and sufficiently smooth. However, we have seen that relaxing the design problem requires introduction of "generalized plate models" based on rapidly varying thicknesses. Since this appears to violate the hypotheses on which (3.2) is based, it is important to ask:

136

<table>
<tr><td>(Q1)</td><td>What is the relation between the homogenized plate equation (2.2), its relaxed compliance $\tilde{L}$, and solutions of the equations of three-dimensional elasticity?</td></tr>
</table>

This leads naturally to a second, much harder question:

<table>
<tr><td>(Q2)</td><td>Consider linear elasticity on the three dimensional plate domain $|x_3| < \varepsilon\, h_\varepsilon(x_1,x_2)$, $(x_1,x_2) \in \omega$, with vertical load $\varepsilon^3 F$ per unit midplane area. Can one characterize (in a simple way) the lowest limiting point that may be obtained for the compliances, as $\varepsilon \to 0$, by letting h_ε vary with ε subject to $h_{min} < h_\varepsilon < h_{max}$, $\int_\omega h_\varepsilon = c$?</td></tr>
</table>

These issues are the focus of our recent work [13,14,15] which will be summarized in the next two sections. Our answer to the first question is that (2.2) correctly models plates whose thickness varies on a length scale that is large compared with the mean thickness but small compared with the plate diameter. We do not have a complete answer to (Q2); however, we are able to compute the effect of thickness variation on a length scale comparable to or shorter than the mean thickness. These lead to different results than those predicted by the Kirchhoff theory.

4. Summary of our Generalized Plate Models

The first important step towards answering the questions Q1 and Q2 is to understand the influence of rapid thickness variation on the (rescaled) compliance of a "three-dimensional plate",

$$\frac{1}{2} \int_{\partial_\pm \Omega_\varepsilon} F |n_3^\varepsilon| u_3^\varepsilon = \varepsilon^{-3} \left(\varepsilon^3 \int_{\partial_\pm \Omega_\varepsilon} F |n_3^\varepsilon| u_3^\varepsilon - \int_{\Omega_\varepsilon} \sigma(\underline{u}^\varepsilon) \cdot e(\underline{u}^\varepsilon) \right),$$

where $\partial_\pm \Omega_\varepsilon$ denotes the upper and lower surfaces of the plate. We shall assume that the thickness is locally periodic with period $\delta = \delta(\varepsilon)$; in other words

(4.1)
$$\Omega_\varepsilon = \{\underset{\sim}{x}, x_3): \underset{\sim}{x} \in \omega, \quad |x_3| < \varepsilon H(\underset{\sim}{x}; \underset{\sim}{x}/\delta)\} ,$$

where $H(\underset{\sim}{x}; \underset{\sim}{n})$ is periodic in its second variable with period 1. Though the dependence of δ on ε could in principle be arbitrary, it is convenient to choose

(4.2)
$$\delta = \varepsilon^a, \quad 0 < a < \infty .$$

We are convinced that (4.1) and (4.2) represent no loss of generality toward the problem of characterizing the limiting compliances.

The asymptotics of $\underline{u}^\varepsilon$ as $\varepsilon \to 0$ can be studied using the method of multiple scales [13,14]. (A closely related problem, involving smoothly varying thickness but rapidly varying elastic moduli, was analyzed simultaneously and independently by Caillerie [8].) The main conclusion is that the energy of the elastic displacement $\underline{u}^\varepsilon$ approaches that of a fourth-order equation

$$\frac{\partial^2}{\partial x_\alpha \partial x_\beta} (M_{\alpha\beta\gamma\delta} \frac{\partial^2 w}{\partial x_\gamma \partial x_\delta}) = F \quad \text{in} \quad \omega$$

(4.3)

$$w = \frac{\partial w}{\partial n} = 0 \quad \text{on} \quad \partial\omega$$

Since the compliance is -2 times the energy, the compliances also converge. The solution w of (4.3) is the limiting vertical displacement; therefore $M_{\alpha\beta\gamma\delta}$ represents an <u>effective rigidity tensor</u>, relating midplane curvature to bending moments in the limit as $\varepsilon \to 0$. It depends not only on $H(\underset{\sim}{x}; \underset{\sim}{n})$ - corresponding to the geometry of the variation - but also on the choice of scaling, specifically on whether $a < 1$, $a = 1$, or $a > 1$. Formulae are given in [13] for each of the rigidities $M^{a<1}$, $M^{a=1}$, and $M^{a>1}$, in terms of the solutions of certain periodic boundary-value problems.

The tensor $M^{a<1}$ has a simple interpretation: it is precisely what one obtains by homogenizing the plate equation (1.1). Thus for "one family of stiffeners" and an isotropic elastic law, $M^{a<1}$ is given by (2.3). In particular, this analysis yields an answer to our first question (Q1):

<u>Rapid variation of the thickness of the "three-dimensional plate" is correctly modelled by homogenizing the Kirchhoff plate equation (only) if it occurs on a length scale larger than the average thickness.</u> (A1)

There is an equally simple interpretation for $M^{a>1}$, involving homogenization of a rough boundary. Consider the system of elasticity on the domain (4.1) with ε fixed, as $\delta \to 0$. The limiting displacement will solve a new elasticity problem on the "smoothed" plate domain

$$\{(\underset{\sim}{x},\ x_3): \underset{\sim}{x} \in \omega,\quad |x_3| < \varepsilon \max_{\underset{\sim}{n}} H(\underset{\sim}{x};\underset{\sim}{n})\}$$

with non-constant, effective elastic moduli $B'_{ijk\ell}(\underset{\sim}{x})$ (see the appendix of [13] for details). Applying Kirchhoff plate theory to this "smoothed" structure corresponds to taking the limit $\varepsilon \to 0$, and it yields precisely the rigidity tensor $M^{a>1}$. (This calculation remains at present merely formal. The homogenization of rough boundaries was made rigorous for a scalar equation in [7], but the system of elasticity presents additional difficulties.)

The intermediate case $a = 1$ has, unfortunately, no such simple interpretation. The tensor $M^{a=1}$ depends on auxiliary functions $\underset{\sim}{\phi}^{\alpha\beta}(\underset{\sim}{x};\underset{\sim}{n})$, obtained by solving the elastotatic boundary value problem

$$\frac{\partial}{\partial n_j}\ \Sigma_{ij}\ (\underset{\sim}{\phi}^{\alpha\beta}) = 0 \quad\text{in}\quad Q(\underset{\sim}{x})$$

(4.4)
$$\Sigma_{ij}(\underset{\sim}{\phi}^{\alpha\beta})\nu_j = \begin{cases} n_3 \tilde{B}_{i\gamma\alpha\beta}\nu_\gamma & i=1,2 \\ 0 & i=3 \end{cases} \quad\text{on}\quad \partial_{\pm}Q(\underset{\sim}{x})$$

$$\underset{\sim}{\phi}^{\alpha\beta} \text{ periodic in } \underset{\sim}{n}\ .$$

Here $Q(\underset{\sim}{x})$ denotes the rescaled periodically varying "plate"

$$Q(\underset{\sim}{x}) = \{\ \underset{\sim}{n}\ :\ |n_3| < H(\underset{\sim}{x};\underset{\sim}{n})\}\ ,$$

$\Sigma_{ij}(\underset{\sim}{\phi})$ is the stress $B_{ijk\ell}E_{k\ell}(\underset{\sim}{\phi})$ associated to the strain $E_{k\ell}(\underset{\sim}{\phi}) = \frac{1}{2}(\partial\phi_k/\partial n_\ell + \partial\phi_\ell/\partial n_k)$ and $\underset{\sim}{\nu}$ is the outward normal to Q; notice that $\underset{\sim}{x}$ enters (4.4) only as a parameter. The formula for $M^{a=1}$ is

$$(4.5) \qquad M^{a=1}_{\alpha\beta\gamma\delta} = \frac{2}{3}\,\overline{H^3}\,\tilde{B}_{\alpha\beta\gamma\delta} - \overline{\int_{-H}^{H} \Sigma_{ij}(\underline{\phi}^{\alpha\beta})E_{ij}(\underline{\phi}^{\gamma\delta})\,dn_3}\,,$$

where as usual the overbar denotes an average over $\underset{\sim}{n} = (n_1,n_2)$. In most cases the actual calculation of $M^{a=1}$ must be done numerically, by solving a finite element approximation of (4.4).

There is a sense in which the $a = 1$ model includes the other two. For any $\lambda > 0$, one can apply (4.4) - (4.5) to the λ-periodic function $H_\lambda(\underset{\sim}{x};\underset{\sim}{n}) = H(\underset{\sim}{x};\underset{\sim}{n}/\lambda)$; this amounts to taking $\delta = \lambda\epsilon$ in (4.1). The tensor so obtained converges to $M^{a<1}$ (rigorously) as $\lambda \to \infty$, and to $M^{a>1}$ (formally) as $\lambda \to 0$; see [14] for details.

The convergence of three-dimensional elastotatics to our generalized plate model has been proved with mathematical rigor only for the intermediate case $a = 1$. In [14] we considered a sufficiently regular $H = H(\underset{\sim}{n})$, taken for simplicity not to depend on the "slow" variable $\underset{\sim}{x}$. We established, among other things, that the relative energy norm error is of order $\sqrt{\epsilon}$ as $\epsilon \to 0$. The main ingredients of the proof are a pair of integral estimates, an averaging lemma, and lots of integration by parts. The first integral estimate is a version of Korn's inequality for thin domains with mean thickness ϵ and thickness variation on length scale $\geqslant \epsilon$, making explicit the dependence of the "constant" on ϵ. The second inequality asserts a weak form of Kirchhoff's hypothesis for the elastostatic displacement $\underline{u}^\epsilon$, as a consequence of the symmetries of the problem (3.1). The averaging lemmas serve to replace certain rapidly-varying expressions by their mean values; they quantify the rate at which a periodic function converges (weakly) to its mean as the period tends to zero. A convergence proof could probably be given for the $a<1$ case using similar methods, but new ideas seem required to handle $a>1$. A particular stumbling block is Korn's inequality for Ω_ϵ: as $\epsilon \to 0$, the "constant" blows up faster when $a>1$ than it does for $a< 1$.

5. Comparison of the Models for One Family of Stiffeners

We have seen that three-dimensional elasticity supports several different models for thin plates with rapidly varying thickness, depending on how the length scale of thickness variation compares with the mean thickness. For applications to structural optimization, it is natural to ask: which scaling gives the strongest structure? A rather complete answer is available in the case of plates with "one family of stiffeners in a specified direction."

One way of comparing the "strength" of two plates is to compare the quadratic forms defined by their rigidities M and $\tilde{M}$. We say that M is weaker than $\tilde{M}$ if $M \leqslant \tilde{M}$, in other words if

$$M_{\alpha\beta\gamma\delta} \, t_{\alpha\beta} \, t_{\gamma\delta} \leqslant \tilde{M}_{\alpha\beta\gamma\delta} \, t_{\alpha\beta} \, t_{\gamma\delta}$$

for every symmetric 2×2 tensor $t_{\alpha\beta}$. By the variational characterization (1.4), the weaker plate has the greater compliance under any load. It may happen, of course, that neither plate is weaker than the other; in that case the ordering of the compliances will be load-dependent.

We shall suppose for simplicity that the thickness variation h_ε is periodic and depends on x_1 alone:

$$(5.1) \qquad\qquad h_\varepsilon(\underset{\sim}{x}) = H(x_1/\varepsilon^a).$$

As remarked in section 2, the generalization to slowly varying stiffener geometry or direction - <u>e.g.</u> to axisymmetric plates with circumferential stiffeners - is immediate. Since we wish to compare the different scalings $a < 1$, $a = 1$, and $a > 1$, it is convenient to fix the choice of H in (5.1). The corresponding effective rigidities $M^{a<1}$, $M^{a=1}$, and $M^{a>1}$ are as discussed in section 4; notice that all three plates have the same average rescaled thickness $\bar{h}$

Surprisingly, the ordering of the strengths of these structures depends on the three-dimensional elastic material used to make them. For plates made from an isotropic elastic material using one family of stiffeners, the $a > 1$ scaling is weakest and the $a < 1$ scaling strongest:

(5.2) $\qquad M^{a>1} < M^{a=1} < M^{a<1}$ <u>for an isotropic elastic law.</u>

The left inequality in (5.2) holds even for anisotropic materials:

(5.3) $\qquad M^{a>1} < M^{a=1}$ <u>in general</u> (for one family of stiffeners).

However, the right inequality in (5.2) can fail for some anisotropic laws and some choices of the profile (H):

(5.4) $\qquad M^{a<1} \not< M^{a=1}$ <u>in general</u> (even for one family of stiffeners).

The proofs, presented in [15], make use of variational principles for each of the three effective rigidity tensors $M^{a<1}$, $M^{a=1}$, and $M^{a>1}$.

More quantitative comparison requires numerical calculation of the effective rigidity tensors. Figure 1 shows the rescaled cross-section (the graph of H) for one of the examples presented in [13]. By cutting along the horizontal midline and glueing together opposite ends of the stiffeners we get a new cross-section with the same amount of material, Figure 2. Tables 1 and 2 list the effective rigidities for these geometries, using an isotropic material with Young's modulus $E = 1.0$ and Poisson's ratio $\nu = 0.25$. The cases $a<1$ and $a>1$ were done using explicit formulas, which are easily derived as in [13]. Calculating $M^{a=1}_{\alpha\beta\gamma\delta}$ requires finding the energies of a pair of two-dimensional cell problems, one involving plane strain and the other antiplane shear. This was done using the FEARS finite element code, developed at the University of Maryland. (Since the cross-section in figure 2 is not of the form (4.1), the analysis of [13,14,15] does not strictly speaking apply. However, the arguments presented there are easily modified to include this case.)

The data in tables 1 and 2 naturally satisfy (5.2). More interesting is the observation that $M^{a=1}$ and $M^{a>1}$ are quite close, while $M^{a<1}_{1212}$ is much greater than $M^{a=1}_{1212}$ in each case. It is not surprising that figure 2 is much stiffer than figure 1; we understand that N. Olhoff and his collaborators are currently studying the use of geometries such as that in figure 2 for compliance optimization.

Our assertion (5.4) concerning the anisotropic case is based on an explicit counterexample: for an elastic law of the form

$$(5.5) \qquad \begin{aligned} B_{iiii} &= \lambda + 2\mu \\ B_{iijj} &= \lambda \qquad i \neq j \\ B_{1212} &= \mu \ , \ B_{1313} = B_{2323} = \mu' \end{aligned}$$

with μ' sufficiently large, and for a thickness profile of the form

$$(5.6) \qquad H(\eta_1) = h_0(1 + \sigma \cdot \cos 2\pi \eta_1)$$

with h_0 and σ sufficiently small, we showed that

$$(5.7) \qquad M_{1111}^{a<1} < M_{1111}^{a=1} \quad \text{and} \quad M_{2222}^{a<1} < M_{2222}^{a=1} .$$

Since the inequality

$$M_{1212}^{a=1} < M_{1212}^{a<1}$$

holds for the elastic law (5.5) (and more generally whenever $\widetilde{B}_{1112} = 0$), this gives an example in which $M^{a=1} - M^{a<1}$ is indefinite. We note that (5.5) is only a minor modification of the isotropic law, which corresponds to the choices

$$(5.8) \qquad \lambda = \frac{E\nu}{(1+\nu)(1-2\nu)} \quad \mu = \mu' = \frac{E}{2(1+\nu)} .$$

We hoped at first that this would lead to examples of practical significance. If the difference $M_{1111}^{a=1} - M_{1111}^{a<1}$ were large, then use of the $a = 1$ plate (with stiffeners in the direction of greatest bending) could be advantageous for some design problems. However, practical numerical examples of (5.7) are exceedingly hard to find. For the geometries of figures 1 and 2, our calucluations give $M^{a=1} < M^{a<1}$ for the elastic law (5.5). The best example we found - and it's not a good one - is shown in Figure 3. With λ and μ chosen by (5.8) with $E = 1.0$ and $\nu = 0.25$, one obtains easily that

$$(5.9) \qquad M^{a<1} = . \ 192901 \times 10^{-2} .$$

(The $a<1$ model does not depend on the value of μ' .) On the other hand

$$(5.10) \qquad M_{1111}^{a=1} = \frac{64}{45} \int_0^{1/2} (\frac{5}{32} - \frac{1}{16} \ t)^3 dt - \mathcal{E}_0 ,$$

where $\mathcal{E_{\varphi}}$ is the strain energy of the cell problem (4.4) - which depends, of course, on μ' . For the isotropic material with $E = 1.0$ and $\nu = 0.25$, μ' is equal to 0.4; for the calculations reported here we took $\mu' = 5.0$. (Larger values of μ' did not increase the significance of the findings, but only served to increase the numerical error.) The value of $\mathcal{E_{\varphi}}$ computed by FEARS was 0.7119×10^{-4}, using 574 elements in a subdivision based on just 1/4 of the unit cell. The code has a built-in error estimator, which says in this case that the energy is off by at most 0.95%. Extensive practical experience with FEARS indicates that the true error exceeds the estimated error by at most 75%. We are thus convinced that the correct value of $\mathcal{E_{\varphi}}$ in (5.10) is bounded above by $.7244 \times 10^{-4}$, which yields

$$M_{1111}^{a=1} > 0.192951 \times 10^{-2} .$$

In view of (5.9), this gives a numerical example of (5.7) - but hardly one of any practical significance!

The implications of these results for structural optimization are clear: for the design of plates made from an isotropic materials using one family of stiffeners with a specified direction, attention may be restricted to the $a < 1$ model (obtained by homogenizing the Kirchhoff plate equation). For plates made from an anisotropic material this is not true, but our numerical experimentation suggests that even so little will be lost in practice by a restriction to Kirchhoff theory.

Care is advised in extending these conclusions to more general situations, such as plates with two or more families to stiffeners. We proved in [15] that $M^{a<1}$ is strongest and $M^{a>1}$ weakest if the elastic law satisfies $B_{\alpha\beta33} = 0$ — which includes the isotropic law with Poisson's ratio ν set equal to zero. But in general the relative strengths appear to depend on both the elastic moduli and the form of the thickness variation.

144

6. Directions for the Future

We indicate some of the areas in which further work is needed.

a) The correctness of our $a > 1$ model as a limit of three-dimensional elasticity has yet to be proved.

b) We still hope for an example (perhaps using two families of stiffeners) where the $a=1$ model leads to a significantly stronger structure than does homogenization of the Kirchhoff theory.

c) Section 2 described a relaxation of the compliance optimization problem for Kirchhoff theory with "one family of stiffeners in a specified direction." It remains to prove an existence theorem, _i.e._ to establish that this is the full relaxation. The corresponding problem without the restriction to "one family of stiffeners" is more difficult. A relaxation (partial or full?) has been proposed in [23]; we understand that Gibianski and Cherkaev also have made progress in this direction.

d) We have shown that Kirchhoff theory suffices for the optimization of plates made from an isotropic material with one family of stiffeners. This result seems likely to fail, however, for more general geometries. Therefore the best relaxation would be one based not on Kirchhoff theory but instead on three-dimensional elasticity - _i.e._ on our generalized plated models.

e) We have discussed only the compliance optimization of linearly elastic plates. In plasticity, the analogous problem is to maximize the limit multiplier of a given load. Rapid thickness variation arises naturally in that context, too, and relaxed formulations have been proposed in [22,25]. It remains, however, to prove that the models used there correctly represent the asymptotic behavior of three dimensional plastic structures. It also remains to check whether a full relaxation has been achieved.

This research was partially supported by NSF grant DMS-8312229 (RVK), ONR grants N00014-83-K-0536 (RVK) and N00014-85-K-0169 (MV), the Sloan Foundation, and the Institute for Mathematics and its Applications.

References

[1]   J.-L. Armand and B. Lodier, "Optimal design of bending elements", Int. J. Num. Meth. Engng., 13, 1978, pp. 373-384.

[2]   J.-L. Armand, K.A. Lurie, and A.V. Cherkaev, "Optimal control theory and structural design," in Optimum Structural Design, Vol. 2, R.H. Gallagher, E. Atrek, K. Ragsdell, and O.C. Zienkiewicz, eds., J. Wiley and Sons, 1983.

[3]   N.V. Banichuk, Problems and methods of optimal structural design. Plenum Press, New York and London, 1983.

[4]   M. Bendsøe, These proceedings.

[5]   M.P. Bendsøe, "On obtaining a solution to optimization problems for solid, elastic plates by restriction of the design space." J. Struct. Mech., 11, (1983/84), pp. 501-521.

[6]   A. Bensoussan, Lions J.-L. and Papanicolaou, G., Asymptotic analysis for periodic structures, North-Holland, Amsterdam, 1978.

[7]   R. Brizzi and J.P. Chalot, Homogénéisation de frontiére. Thése, Université de Nice, 1978.

[8]   D. Caillerie, "Thin elastic and periodic plates", Math. Methods Applied Sci. 6, 1984, pp. 159-191.

[9]   K.-T. Cheng and N. Olhoff, "An investigation concerning optimal design of solid elastic plates," Int. J. Solids Structures, 17, 1981, pp. 305-323.

[10] K.-T. Cheng and N. Olhoff, "Regularized formulation for optimal design of axisymmetric plates." Int. J. Solids Structrues, 18, 1982, pp. 153-169.

[11] P.G. Ciarlet and P. Destuynder, "A justification of the two-dimensional linear plate model" J. Mecanique, 18, 1979, pp. 315-344.

[12] R.V. Kohn and G. Strang, "Optimal design and relaxation of variational problems," to appear in Comm. Pure Appl. Math.

[13] R.V. Kohn and M. Vogelius, "A new model for thin plates with rapidly varying thickness." Int. J. Solids Structures, 20, 1984, pp. 333-350.

[14] R.V. Kohn and M. Vogelius, "A new model for thin plates with rapidly varying thickness II: a convergence proof." Quart. Appl. Math., 43, 1985, pp. 1-22.

[15] R.V. Kohn and M. Vogelius, "A new model for thin plates with rapidly varying thickness III: Comparison of different scalings." Quart. Appl. Math., 43, 1985, to appear.

[16] K.A. Lurie and A.V. Cherkaev, "G-closure of some particular sets of admissible material characteristics for the problem of bending of thin plates," J. Opt. Th. Appl.,1984, pp. 305-316.

[17] K.A. Lurie, A.V. Fedorov and A.V. Cherkaev, "On the existence of solutions to some problems of optimal design for bars and plates.", J. Opt. Theory Appl. 42, 1984, pp. 247-281.

[18] D. Morgenstern, "Herleitung der plattentheorie aus der dreidimensionalen elastizitäts-theorie." Arch. Rational Mech. Anal., 4, 1959, pp. 145-152.

[19] F. Murat and L. Tartar, "Calcul des variations et homogénéisation," in Les Methodes de ℓ'Homogénéisation: Theorie et Applications en Physique, Coll. de la Dir. des Etudes et Recherches de Electricité de France, Eyrolles, Paris, 1985, pp. 319-370.

[20] N. Olhoff and J.E. Taylor, "On structural optimization", J. Appl. Mech., 50, 1983, pp. 1134-1151.

[21] U.E. Raitum, "On optimal control problems for linear elliptic equations," Soviet Math. Dokl. 20, 1979, pp. 129-132.

[22] G.I.N. Rozvany, N. Olhoff, K.-T. Cheng, and J. Taylor, "On the solid plate paradox in structural optimization," J. Struct. Mech., 10, 1982, pp. 1-32.

[23] G.I.N. Rozvany, T.G. Ong, R. Sandler, W.T. Szeto, N. Olhoff, and M.P. Bendsøe, "Least-weight design of perforated plates," preprint.

[24] E. Sanchez-Palencia, Non-homogeneous media and vibration theory, Lecture Notes in Physics, 127, Springer-Verlag, 1980.

[25] C.-M. Wang, G.I.N. Rozvany, and N. Olhoff, Optimal plastic design of axi-symmetric solid plates with a maximum thickness constraint, Computers & Structures, 18, 1984, pp. 653-665.

	a<1	a=1	a>1
M_{1111}	.015	.012	.011
M_{1122}	.004	.003	.003
M_{2222}	.334	.334	.334
M_{1212}	.113	.006	.004

Table 1: Effective rigidities for figure 1.

	a<1	a=1	a>1
M_{1111}	.777	.687	.678
M_{1122}	.194	.172	.169
M_{2222}	.851	.845	.844
M_{1212}	.321	.262	.254

Table 2: Effective rigidities for figure 2.

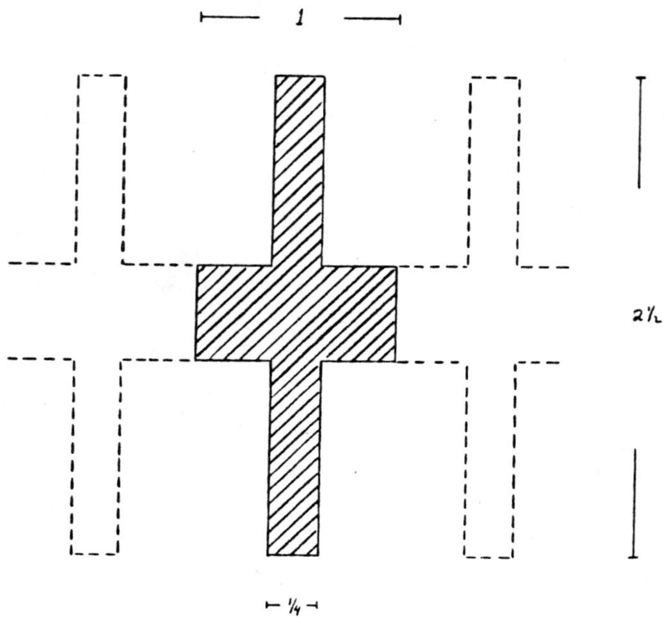

Figure 1

Figure 2

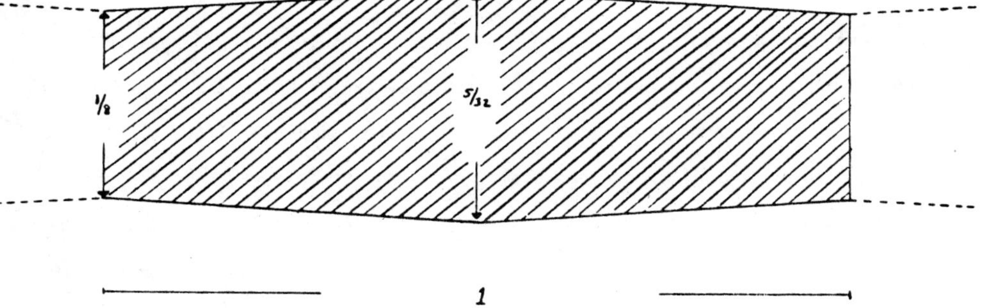

Figure 3

MODELLING THE PROPERTIES OF COMPOSITES BY LAMINATES

by

G.W. Milton

California Institute of Technology, 405-47
Pasadena, CA 91125

Laminate materials of the type introduced by Schulgasser [1] and Bruggeman [66,67] are studied and the extent to which they can simulate the transport properties of other composites is explored. Laminates with chirality, with especially high or low field concentrations, or which attain various bounds are constructed. The Hashin-Shtrikman bounds on the shear modulus are demonstrated to be optimal, being attained by a hierarchical laminate material. While the conductivity function of two-component composites can be simulated by laminates, an example suggests this does not extend to five-component composites. Attention is drawn to the connection between conductivity functions, Stieltjes functions, and bounds.

1. Introduction

It was first recognized by Bruggeman [66, 67] that the effective conductivity can be calculated _exactly_ for a wide class of composites constructed via a laminating procedure. These laminate materials have inhomogeneities on multiple length scales: typically one begins by slicing two components and placing the slices in alternate order to form a multilayered sandwich, called a laminate of rank 1. The sandwich, in turn, is sliced in a different direction (on a _much_ larger length scale) and combined with, say, slices of another multilayered sandwich to produce a more complex laminate of rank 2, such as sketched in Fig. 1. This process can be continued to produce laminates of arbitrarily high rank, and thereby a tremendous variety of composites can, in principle, be constructed. Of course, laminates of high rank are difficult to manufacture because they are structured on many widely separated length scales. For this reason they are more useful as a theoretical tool to gain insight into the transport properties of composites than of direct practical importance. The

transport properties of these materials are easy to evaluate since the local
fields are piecewise uniform. Several concise equations for calculating the
properties of laminates have been formulated by Backus [2] and by Tartar [3].

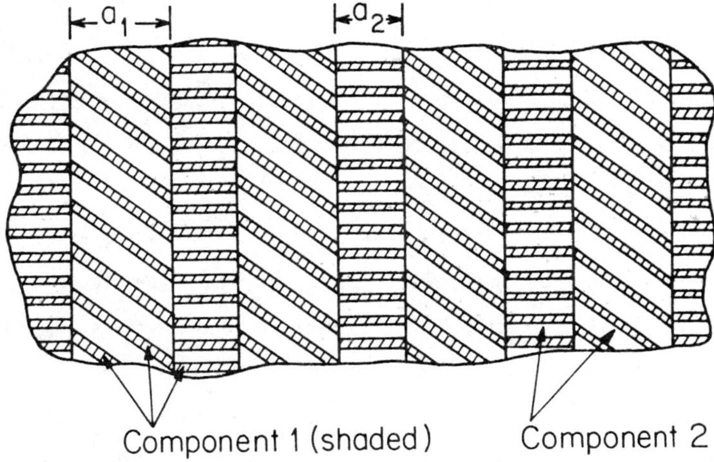

Fig. 1. Sketch of a typical two-dimensional, two-component, laminate material of
rank 2. The widths a_1 and a_2 of the composite slabs should be much smaller
than the thickness of the layers in each slab.

Laminates of non-trivial rank became the focus of increasing attention
after the pioneering work of Schulgasser. [1,4]. He was interested in the
problem of finding the maximum range of effective conductivity that an isotropic
polycrystal, composed of grains each sharing the same principle conductivities
σ_a, σ_b and σ_c , can take as the orientations, shapes and configurations of the
grains are varied. Using an ingeneous construction scheme, Schulgasser found that
the well known bound

$$\sigma^* \leqslant \frac{1}{3} \left(\sigma_a + \sigma_b + \sigma_c \right) \qquad (1.1)$$

on the effective conductivity, σ^* , of a three dimensional isotropic polycrystal is
in fact attained by a laminate of rank 3. It is still not known what isotropic
polycrystal has the lowest effective conductivity although Schulgasser [4,5] and
Lurie and Cherkaev [6] have made notable progress.

Following Schulgasser's earlier work, I discovered [7,8] that laminate materials are important in attaining many bounds on the complex dielectric constant of a two-component composite. In totally independent work, Lurie and Cherkaev [9] found that laminates of rank 2 attain bounds, known as G-closures, correlating the eigenvalues of the effective conductivity tensor of two-dimensional, two-component composites. This result was subsequently extended to arbitrarily high dimension, by Murat and Tartar [10] and by Lurie and Cherkaev [11]. Their bounds, which encompass the Hashin-Shtrikman bounds [12] for isotropic composites, are described in Section 2.

The outline of this paper is as follows. First I will give a brief review of some exciting recent work on G-closures, i.e. on optimal bounds for the effective conductivity (or elasticity) tensor. A simple physical argument is given that explains why laminate materials attain elementary G-closures. The argument is applied to show that the the Hashin-Shtrikman bounds on the shear modulus [13] are, in fact, attained by a hierarchical laminate, basically similar to a model first introduced by Sen, Scala and Cohen [14]. This demonstrates the optimality of the Hashin-Shtrikman bounds on the shear modulus.

Of particular interest amongst the recent work on laminates is Lurie and Cherkaev's conjecture [15] that the conductivity tensor (or elasticity tensor) of an composite with specified volume fractions can be modelled by an appropriate laminate material with the same volume fractions of the components. In Section 3, a natural generalization of their conjecture is examined, namely the question whether the functional dependence of the effective conductivity on the component conductivities can always be modelled by a laminate. The present work indicates that while the generalized conjecture may be true for two-component composites, it is unlikely to extend to composites with an arbitrary number of components. In the course of this discussion the analytic properties of the effective conductivity are reviewed and the relationship with Stieltjes functions is discussed. Attention is drawn to the connection between established bounds on composites and the bounds on Stieltjes functions derived by Henrici and Pfluger [63], Common [53], and Baker [16], among others.

The question of what geometries produce the highest and lowest field concentrations is considered in Section 4. This problem dates back to early work of Beran [17] who derived some elementary bounds on the variance of the electric field. Here _threshold exponents_ are defined to provide a crude, but useful, measure of field concentrations. By comparing threshold exponents it is found that laminates can produce higher (and lower) field concentrations than the field concentrations that exist near sharp corners.

My goal throughout is to demonstrate that laminates represent an important class of composite materials, which exhibit a wide range of transport properties. I believe laminates are destined to prove an important theoretical tool for modelling the properties of two-component composites, and to a lesser extent, multicomponent composites. The exact limitations of the utility of laminate materials still need to be explored.

2. G-Closures and the Conjecture of Lurie and Cherkaev.

Given a set U of conductivity (or elasticity) tensors, corresponding to various components, the G-closure of U, GU, is defined as the complete set of effective conductivity tensors (or elasticity tensors) associated with composites formed from these components. As Kohn and Strang [15] and Bendsøe [18] have demonstrated in several illustrative examples, G-closures are of fundamental importance in solving optimization problems. The G-closures of many sets have been determined by Tartar and Murat [3,10] and Lurie and Cherkaev [6,9,11,19,20] in a series of outstanding papers. The simplest example is when U consists of a single tensor with real eigenvalues σ_a and σ_b corresponding to the principal conductivities of a perfect two-dimensional crystal. Then GU consists of those tensors with eigenvalues σ_x^*, σ_y^* satisfying,

$$\sigma_x^* \, \sigma_y^* = \sigma_a \, \sigma_b \, , \quad \sigma_a < \sigma_x^* < \sigma_b \, , \qquad (2.1)$$

which thus describes an arc of a hyperbola in the (σ_x^*, σ_y^*) plane. This result, first explicitly stated by Lurie and Cherkaev [19,20] dates back, in part, to earlier work by Dykhne [21], Mendelson [22] and Schulgasser [4] and is based on a duality relationship due to Keller [23].

When the volume fractions $\underline{f}$ of the components are specified a different type of G-closure is needed: the set $G_f U$ is defined as the family of effective conductivity tensors of composites that can be formed from the components in the set U, in the prescribed proportions, $\underline{f}$. For example, suppose the set U consists of two isotropic tensors $\sigma_1\underline{I}$, $\sigma_2\underline{I}$, representing two isotropic components in a space of dimension d, in proportions f_1 and f_2 with say $\sigma_1 > \sigma_2$. To describe $G_f U$ we introduce, on the set of real symmetric d-dimensional matrices, the transformation

$$\Lambda_f\ (\underline{\sigma}) \equiv (\langle\sigma\rangle\underline{I} - \underline{\sigma})(\underline{\sigma}\langle 1/\sigma\rangle - \underline{I})^{-1} , \tag{2.2}$$

in which $\underline{I}$ is the identity matrix and

$$\langle\sigma\rangle \equiv f_1\sigma_1 + f_2\sigma_2 , \quad \langle 1/\sigma\rangle \equiv f_1/\sigma_1 + f_2/\sigma_2 \tag{2.3}$$

are the mean conductivity and mean resistivity. Murat and Tartar [10] and Lurie and Cherkaev [9,11] found the set $\Lambda_f(G_f U)$ is __independent__ of f and consists of those tensors with positive eigenvalues λ_i^* (i = 1,2,...d) satisfying the inequalities

$$\sum_{i=1}^{d} (1 + \sigma_1/\lambda_i^*)^{-1} < 1 < \sum_{i=1}^{d} (1 + \sigma_2/\lambda_i^*)^{-1} . \tag{2.4}$$

These bounds, first proved using the method of __compensated__ __compactness__ developed by Tartar and Murat (see Tartar [24]) can also be established from the Hashin-Shtrikman variational principles [25] or from the analytic properties of the effective conductivity tensor $\underline{\sigma}^*(\sigma_1,\sigma_2)$ as a function of σ_1 and σ_2 [26]. The transformation Λ_f is significant becuase it preserves many of the special analytic properties of $\underline{\sigma}^*(\sigma_1,\sigma_2)$ described in section 3.

The bounds defined by (2.4) are optimal and reduce to the well-known Hashin-Shtrikman bounds [12] for isotropic composites. Tartar and Murat found that any point on the boundary of $G_f U$ corresponds to an aggregate consisting of aligned coated confocal ellipsoids, of various sizes, filling all space. These composites, also associated with bounds on the complex dielectric constant [7,8,27], represent a natural generalization of the Hashin-Shtrikman coated sphere geometries [12].

Laminate materials of rank d, or higher, also serve to attain the bounds (2.4) and thus can be used to model the properties of coated ellipsoid geometries. The physical basis for this important result of Lurie and Cherkaev [9,11] and Tartar [3] can be understood from the recent work of Kohn and myself [25]. In addition to generalizing (2.4) to the elasticity case, thereby obtaining a comprehensive set of bounds on the effective elasticity tensor of anisotropic multicomponent media, we found the Hashin-Shtrikman variational principles [12] imply that the lower, or upper, bound in (2.4) is attained if and only if the field in component 1, or component 2, is (almost everywhere) uniform. Thus any laminate material that is constructed by starting with a matrix of component 1 (or 2) and successively laminating it with the other component, until a material of some desired rank is achieved, must necesarily attain the bounds (2.4) since the field in the starting matrix material is always uniform when the applied field is uniform.

This same argument can be applied to resolve an outstanding problem. Ever since Hashin and Shtrikman first formulated their bounds [13] on the effective shear moduli of two-component isotropic composites, the question of whether these bounds could be attained remained unanswered. Hashin and Shtrikman [12,13] had shown that their bounds on the conductivity and bulk modulus of two component composites were attainable by coated sphere geometries. Later this finding was extended to multicomponent composites, over a restricted range of volume fractions [28]. However the argument used to establish these results breaks down in the shear modulus case since the shear field outside a coated sphere embedded in the appropriate effective medium is not uniform, whereas the pressure field and the electric fields are uniform. This was convincingly demonstrated by Hashin and Rosen [29] and Christensen and Lo [30], among others.

Now consider the hierarchial model of Fig. 2. It is a limiting example of a self-similar model with plate-like grains first introduced by Sen, Scala and Cohen [14] and is obtained by successively laminating a matrix of component 2, with very thin, widely separated plates of component 1 orientated in randomly chosen directions at different levels in the hierarchy. The hierarchy is continued until the desired volume fraction of the components is reached. The transport properties of this material are clearly isotropic in the limit in which an infinitesimal fraction of slabs (or plates) of component 1, is introduced at each level in the hierarchy. Furthermore the shear field in component 2, the starting material, is uniform. Thus the condition for attainability of the Hashin-Shtrikman bounds is met and consequently the hierarchical model of Fig. 2 represents the stiffest isotropic material that can be constructed from two compounds with bulk moduli $\kappa_1 > \kappa_2$ and shear moduli $\mu_1 > \mu_2$. The most compliant isotropic composite is obtained by reversing the roles of the two phases in this construction. When $\kappa_1 - \kappa_2$ and $\mu_1 - \mu_2$ have opposite signs, it is still not known what geometries give the stiffest or most compliant material. I doubt that the best currently available bounds, dervied by Phan-Thien and myself [31], are optimal.

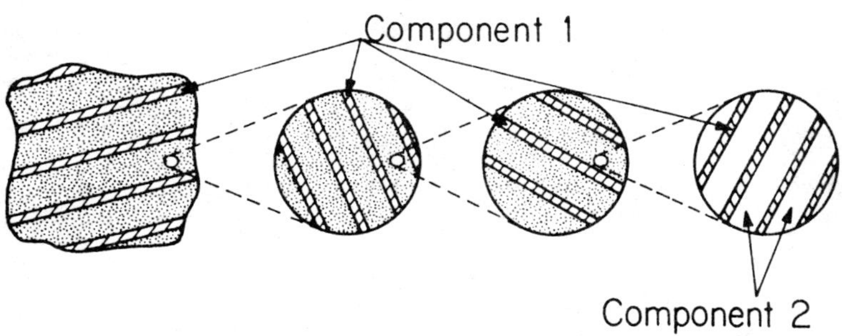

Fig. 2. The first stages in the construction of a hierarchical laminate, corresponding to the stiffest materials that can be manufactured from two components, 1 and 2, with bulk moduli $\kappa_1 > \kappa_2$, and shear moduli $\mu_1 > \mu_2$, in fixed proportions f_1 and f_2. At each stage in the construction an infinitesimal volume fraction of component 1 is introduced, until the desired volume fraction, f_1, of component 1 is reached. The essential geometric features are the same as in an earlier self-similar model of Sen, Scala and Cohen [14], except that here successive length scales must be widely separated. The bulk and shear moduli for this material can be calculated exactly using the iterated dilute limit approximation, also known as the differential scheme: see Norris [32] and references therein.

In totally independent work, Norris [32] and Lurie and Cherkaev [33] also arrived at the above conclusion, finding, through detailed calculations, that the Hashin-Shtrikman bounds on the shear modulus were attained by geometries basically similar to the model in Fig. 2. Also, in independent work presented at this conference, F. Murat, L. Tartar, and G. Francfort found an elegant realization of the Hashin-Shtrikman shear modulus bounds using laminates of _finite_ rank [65]. Clearly laminate materials play an important role in attaining many bounds on the transport properties of composites.

Observations such as these led Lurie and Cherkaev to make a bold conjecture, implicit in one of their recent papers [6]. According to Kohn and Strang [15], "Lurie and Cherkaev have conjectured that sequentially layered composites (laminates) will always suffice to construct the G_f - closure of any set of materials". This conjecture, if true, would be a major breakthrough towards solving the currently intractable problem of determining the G_f closure of an arbitrary set. For this reason alone, the conjecture deserves attention.

To shed some light on the hypothesis of Lurie and Cherkaev, let us now investigate whether the functional dependence of the effective conductivity can be modelled by laminates. If true, this would serve to establish their conjecture.

3. Modelling the Functional Dependence of the Effective Conductivity by Laminates.

The analytic dependence of the effective conductivity $\sigma_{jj}^*(\sigma_1, \sigma_2, \sigma_3, \ldots)$ in a fixed direction x_j as a function of the component scalar conductivities $\sigma_1, \sigma_2, \sigma_3, \ldots$ has been explored by Bergman [34,27] for two-component composites and by Golden and Papanicolaou [35] for multicomponent composites. For a two-component composite, the _homogeneity_ property

$$\sigma_{jj}^*(c\sigma_1, c\sigma_2) = c\sigma_{jj}^*(\sigma_1, \sigma_2), \qquad \text{for all } c, \qquad (3.1)$$

of the effective conductivity, implies that we can set $\sigma_2 = 1$ and study the properties of the single variable conductivity function,

$$q(\sigma_1) \equiv \sigma_{jj}^*(\sigma_1)/\sigma_1 \ , \qquad\qquad (3.2)$$

without any loss of generality. Bergman [34] found that this conductivity function $g(\sigma_1)$ has some remarkably simple and beautiful analytic properties: it can be approximated by a rational function of σ_1 with poles and zeroes alternating along the negative real axis, starting with a pole nearest (or at) the origin and ending with a zero near (or at) minus infinity.

This result, proved rigorously by Golden and Papanicolaou [36], parallels earlier work of Forster [37], who studied the impedance of two-component electrical network as a function of the impedances of the two components (which could be resistors, capacitors or inductors) and arrived at similar conclusions: see Storer [38] and Baker [39] for details. The connection with Bergman's work is easy to appreciate. As discussed elsewhere [7], and as is suggested by Fig. 3, the electrical transport properties of a slab of composite materials positioned between two conducting plates can be modelled by a cubic resistor network, with lattic spacing smaller than the size of inhomogeneities in the composite. In this sense, Bergman's work represents an extension of Forster's result to the continuum limit.

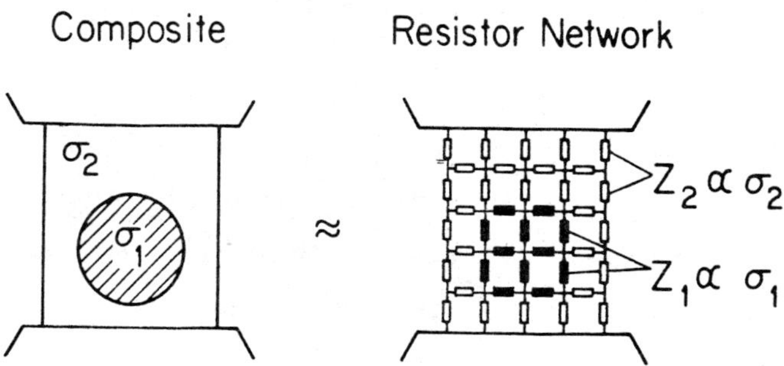

Fig. 3. The transport properties of a slab of composite positioned between two superconducting plates can be simulated by a resistor network of the above type of construction. The impedance of each resistor should be chosen proportional to the resistance at the corresponding point in the composite: this is equivalent to solving the differential equations using a finite difference approach.

The analytic properties of $\underline{\sigma}^*$ (σ_1, σ_2) have many important implications. Bergman [34] discovered that one could use these analytic properties to derive the Hashin-Shtrikman bounds and some other new bounds, when the component conductivities are real. Later, in independent work, Bergman [40,41] and I [7,42] obtained elementary bounds for complex σ_1, σ_2 restricting σ_e to lens-shaped regions of the complex plane. Unfortunately, Bergman at first failed to recognize that my construction of the complex bounds was different in two and three dimensions. Thus, when he found the 3-d bounds for isotropic materials were not attained by a doubly coated sphere geometry, he erroneously concluded [41] that my identification of the 2-d bounds with a doubly coated cylinder geometry was incorrect. A complete proof of both the 2-d and 3-d isotropic complex bounds, and a full discussion of the geometries which attain them, was first given in ref. 7: see also Bergman [27].

Following the derivation of these elementary bounds several infinite nested sequences of optimal bounds were derived [8] and subsequently investigated by McPhedran, McKenzie and Milton [43], McPhedran and Milton [44] and Felderhof [45]: alternative and much simpler derivations of some of these bounds have been given by Milton and Golden [26] and Golden [46], using transformations similar to (2.2). The bounds for real σ_1 and σ_2 coincide with those derived by Beran [47] via variational principles [48]. These bounds have been evaluated to third order, sometimes fourth order, for a variety of realistic disordered materials: see Torquato and Stell [49], Torquato and Beasley [50], Berryman [51], and Felderhof [45] and references therein.

Through private communication with G.A. Baker, Jr. and J.G. Berryman it was recently recognised [26] that the conductivity function $q(\sigma_1)$ is a function that has been extensively studied in the mathematics literature, namely a Stieltjes function. It turns out that most of the important bounds derived by Bergman [27,34,40,41], me [7,8,42], Felderhof [52], and Golden [46], including the Hashin-Shtrikman bounds, could have been directly deduced from known bounds on Stieltjes functions, due to Henrici and Pfluger [63], Common [53] and Baker [16] among others. These bounds are, in fact, closely related to Padé approximants.

For an outstanding and comprehensive review of work on Padé approximants, see
Baker and Graves-Morris [64].

The connection is most apparent between Bergman's approach and Baker's work. As
shown in the appendix, the central fractional linear transformation, used by Bergman
[34], Bergman and Kantor [54] and Golden [46] to derive their bounds for two-component
composites is equivalent to that used earlier by Baker [16]. It is surprising that it
took so long for this simple connection to be recognised, since Stieltjes functions occur in many physical problems [55].

Let us now consider whether the effective conductivity function can be modelled
by laminates. The analytic properties of $q(\sigma_1)$ imply the representation formula

$$\sigma_{jj}^*(\sigma_1,\sigma_2) \approx \sum_{i=1}^{m} A_i \, (q_{1,i}/\sigma_1 + q_{2,i}/\sigma_2)^{-1} \; . \tag{3.3}$$

for the effective conductivity in a fixed direction, where $m \gg 1$ and the geometric
parameters $A_i, q_{1,i}$ and $q_{2,i}$ are real and non-negative, for all i, satisfying

$$\sum_{i=1}^{m} A_i = 1, \; q_{1,i} + q_{2,i} = 1 \; . \tag{3.4}$$

This representation formula clearly implies that the effective conductivity in a
single direction, can be modelled by a laminate material of rank 2, as sketched in Fig.
4: see also appendix B in ref 8. The analogous result for impedance networks is due
to Forster [37], and has proved to be an important tool in the synthesis of electrical
networks [38].

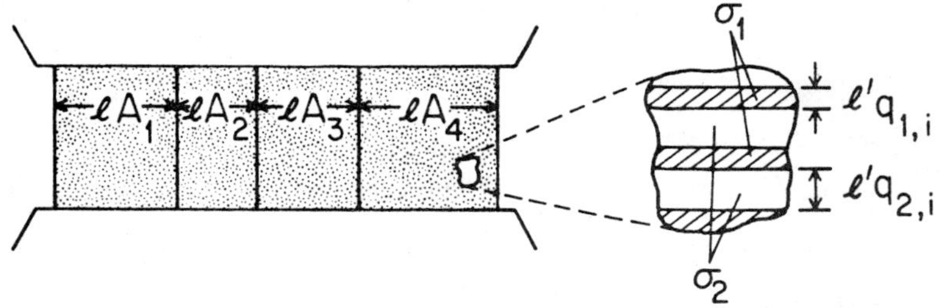

Fig. 4. This type of second rank laminate material serves to model the conductivity
function $\sigma_{jj}^*(\sigma_1,\sigma_2)$ of an arbitrary two-component composite sandwiched between two
superconducting plates, as in Fig. 3. If the composite is two-dimensional and has
reflection symmetry about some line then the laminate has the same response as the composite for all directions of the applied field, i.e. it simulates the tensor function
$\underline{\sigma}^*(\sigma_1,\sigma_2)$.

The question of whether the conductivity tensor function $\underline{\sigma}^*(\sigma_1,\sigma_2)$ can be modelled by laminates remains largely unexplored. It is not difficult to see that the tensor function of any two-component, two-dimensional composite that is invariant under spatial reflection, can be modelled by a rank 2 laminate constructed as in Fig. 4. The invariance under spatial reflection ensures that the eigenvectors of $\underline{\sigma}^*(\sigma_1,\sigma_2)$ do not change direction as the ratio σ_1/σ_2 varies. So we may choose x_1 and x_2 parallel to the eigenvectors of $\underline{\sigma}^*$. Then the interchange relationship,

$$\sigma_{11}^*(\sigma_1,\sigma_2) = \sigma_1\sigma_2/\sigma_{22}^*(\sigma_2,\sigma_1) \tag{3.5}$$

of Keller [23], Dykhne [21] and Mendelson [22], implies that the laminate material of Fig. 4 modelling the function $\sigma_{22}^*(\sigma_1,\sigma_2)$ will also serve to model the remaining eigenvalue $\sigma_{11}^*(\sigma_1,\sigma_2)$ and hence the tensor function $\underline{\sigma}^*(\sigma_1,\sigma)$. Some composites with <u>chirality</u>, i.e. with some degree of left or right-handed asymmetry cannot be modelled by laminates of this construction. Sketched in Figs. 1 and 5 are examples of two such composites, one a laminate material, for which the eigenvectors rotate as the ratio σ_1/σ_2 varies. The simplest way of varying the ratio σ_1/σ_2 , is of course to vary the frequency of the applied field, which thereby provides a test for determining whether a composite has chirality. Homogeneous

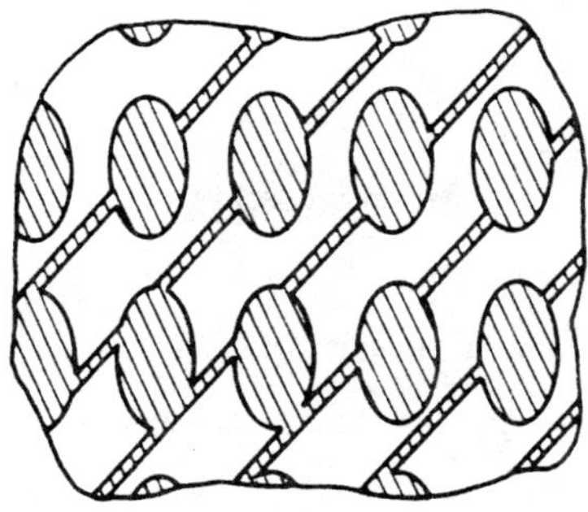

Fig. 5. A composite with chirality. When the conductivity ratio σ_1/σ_2 is nearly unity the eigenvectors of $\underline{\sigma}^*(\sigma_1,\sigma_2)$ align with the axes of the ellipsoids: the filamentary threads of component 1 have little influence on the effective conductivity. As the conductivity ratio is increased the threads carry more current, and thereby cause the eigenvectors to rotate. An example of a laminate materials with chirality is given in Fig. 1.

anisotropic crystals may similarly exhibit chirality to the extent that their eigenvectors are frequency dependent. When these materials are lossy, the eigenstates correspond to states of elliptically polarized radiation. This phenomena, as distinguished from optical activity, is discussed by Born and Wolf [57].

I doubt that the conductivity function $\sigma_{jj}^*(\sigma_1, \sigma_2, \sigma_3, \ldots)$ for multicomponent composites can be modelled by laminates. The analogous result for impedance networks is, in fact, <u>false</u>. For example, the impedance function $Z^*(Z_1, Z_2, Z_3)$ for the Wheatstone bridge of Fig. 6 cannot be modelled by any network consisting of resistors in series-parallel combinations. In the Wheatstone bridge no current flows in the impedance Z_3 when $Z_1 = Z_2$ and consequently $Z^*(Z_1, Z_2, Z_3)$ depends on Z_3 only when $Z_1 \neq Z_2$. By contrast, in any series-parallel network, Z^* depends on Z_3 whenever the impedance Z_3 is present.

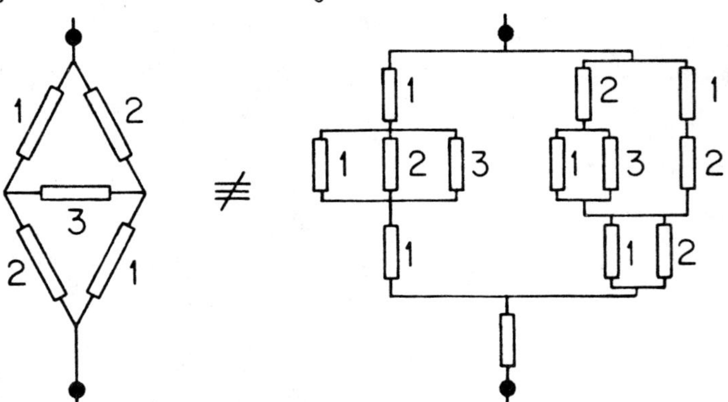

Fig. 6. The impedence function $Z^*(Z_1, Z_2, Z_3)$ of this Wheatstone bridge cannot be modelled by any series-parallel network comprised of linear combinations of the three impedences, Z_1, Z_2 and Z_3 .

The outstanding question is whether there is some three-component composite that is somehow analogous to the Wheatstone bridge. If the effective conductivity $\sigma_{jj}^*(\sigma_1, \sigma_2, \sigma_3)$ of a composite is independent of σ_3 when $\sigma_1 = \sigma_2$, then the volume fraction

$$f_3 = \left. \frac{\partial \sigma_{jj}^*}{\partial \sigma_3} \right|_{\sigma_1 = \sigma_2 = \sigma_3} \tag{3.6}$$

of component 3 must be zero, or at least infinitesimal. So in the search to find a Wheatstone bridge-like composite (constructed from 3 components) we are confronted wi

<u>The</u> <u>Defect</u> <u>Problem</u>: Can the effective conductivity of a multicomponent composite be influenced by an infinitesimal portion of one component when all the constitutents have finite non-zero conductivities? Equivalently, can $\sigma^{*}_{jj}(\sigma_1,\sigma_2,\sigma_3)$ still depend on σ_3 in the limit $f_3 \to 0$ when $\sigma_1,\sigma_2,\sigma_3 \neq 0$ or ∞ ?

While a filamentary thread of superconductor, or a thin wall of insulating material, can dramatically influence the effective conductivity it is hard to imagine a Wheatstone bridge-like composite, where an infinitesimal portion of <u>finitely</u> conducting material channels or diverts a measurable amount of current. Yet it has never been proved impossible to construct such a material and my own attempts to resolve this issue have failed. Kerner [57] and Walpole [58] have suggested it may be possible to construct such a material, while Landauer, [59], I [28] and Lurie and Cherkaev [33] have implied the contrary. The answer to the defect problem depends on the extent to which the current and electric fields can be concentrated in small regions, and this is the focus of Section 4.

Aside from the question of whether a three-component composite can model the properties of a Wheatstone bridge, it is clear from Fig. 7, that a two-dimensional Wheatstone bridge-like composite can indeed be manufactured at the cost of introducing two extra components: one with zero conductivity (σ_4 = 0) and the other

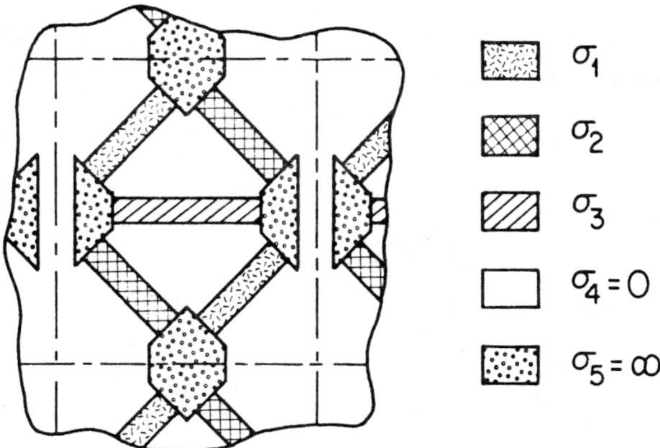

Fig. 7. The unit cell of a periodic five-component composite modelling the conductivity function of a Wheatstone bridge. Of the five components, component 4 (with σ_4 = 0) insulates the circuit elements from each other, while component 5 (with σ_5 = ∞) acts as the solder making electrical contact between the circuit elements.

with infinite conductivity ($\sigma_5 = \infty$). And there appears to be no way the conductivity function of this material, can be simulated by a laminate material.

Field Concentrations, Threshold Exponents and Laminates.

As it is well-known, the electric field, $E(x)$, can be highly concentrated near sharp corners and spikes. Usually one wants to avoid such field concentrations because they lead to dielectric breakdown. Here we search for structures that give especially large field concentrations. Thus the quest is to find the worst possible dielectric material.

We need to have some measure of the degree of field concentration in a composite. So let us define the <u>threshold exponents</u>,

$$\gamma^- \equiv \inf \gamma : \frac{1}{V(\Omega)} \int_\Omega |\underline{E}(\underline{x})|^\gamma \, d\underline{x} < \infty \tag{4.1}$$

$$\gamma^+ \equiv \sup \gamma : \frac{1}{V(\Omega)} \int_\Omega |\underline{E}(\underline{x})|^\gamma \, d\underline{x} < \infty \tag{4.2}$$

in which the integral extends over the macroscopic region Ω , of volume $V(\Omega)$, occupied by the composite. The exponent γ^+ is small only when there are areas with high field concentrations and γ^- is small and negative only when there exist areas in which the electric field is especially feeble. Since the electric field is square integrable, and $|E(x)|^0 = 1$, the threshold exponents necessarily satisfy the inequalities

$$\gamma^- \leqslant 0 \, , \, \gamma^+ \geqslant 2 \, , \tag{4.3}$$

Typically the magnitude of the electric field will be uniformly bounded above and below (by positive constants) and then the exponents are trivially, $\gamma^- = -\infty$ and $\gamma^+ = +\infty$.

Now given two component conductivities, $\sigma_1 > \sigma_2$, let us compare the threshold exponents for a few representative composites built from these components. Near a wedge of component 1 material, with included angle β , surrounded by component 2, the electric potential $\phi(r,\theta)$ in polar co-ordinates r,θ (with $\theta = \pi$ bisecting the wedge) has the asymptotic form

$$\phi(r,\theta) \approx A r^\alpha \cos(\alpha\theta + \psi) \tag{4.4}$$

as $r \to 0$, for $|\theta| < \pi - \frac{1}{2}\beta$, where to satisfy the field equations the exponent α must be a root of

$$\tan \alpha(\pi - \tfrac{1}{2}\beta) = -k \tan (\tfrac{1}{2}\alpha\beta) , \qquad (4.5)$$

in which k is either σ_1/σ_2 (when $\psi = 0$) or σ_2/σ_1 (when $\psi = \pi/2$): see Meixner [60] for details. To obtain an extreme value of α , and hence the best threshold exponents, we choose the wedge angle β such that

$$\cos (\tfrac{1}{2}\alpha\beta) = [\sigma_1/(\sigma_1 + \sigma_2)]^{1/2} \quad \text{or} \quad [\sigma_2/(\sigma_1 + \sigma_2)]^{1/2} , \qquad (4.6)$$

which with (4.5) yields the set of solutions

$$\alpha = m \pm (1/\pi)\sin^{-1}[(\sigma_1 - \sigma_2)/(\sigma_1 + \sigma_2)], \qquad (4.7)$$

where m is any integer $\geqslant 1$. A simple calculation shows that a two-dimensional composite containing many of these wedges, in say the form of diamond-shaped grains orientated both parallel and perpendicular to the applied field, has the threshold exponents

$$\gamma^+ = - \gamma^- = 2\pi\{\sin^{-1}[(\sigma_1 - \sigma_2)/(\sigma_1 + \sigma_2)]\}^{-1} . \qquad (4.8)$$

A checkerboard of touching squares, as studied by Dykhne [21] and Söderberg and Grimvall [61], has even smaller threshold exponents, namely

$$\gamma^+ = - \gamma^- = 2/(1-\alpha'). \qquad (4.9)$$

where

$$\alpha' = (4/\pi)\tan^{-1}(\sigma_2/\sigma_1)^{1/2} \qquad (4.10)$$

is the exponent characterizing the radial dependence of the field singularity near the corners of each square in the checkerboard.

Another model with field singularities is the two dimensional analog of Schulgasser's symmetric material [5]. When the circular grains in the model have laminations alternating in the radial direction the threshold exponents are

$$\gamma^+ = \infty , \quad \gamma^- = 2 - 2(\sigma_1 + \sigma_2)/(\sigma^{1/2}_1 - \sigma^{1/2}_2)^2 \qquad (4.11)$$

and when the laminations alternate in the angular direction, we have

$$\gamma^+ = 2(\sigma_1 + \sigma_2)/(\sigma_1^{1/2} - \sigma_2^{1/2})^2 \ , \ \gamma^- = -\infty \ . \tag{4.12}$$

Physically this makes sense. When the laminations are in the radial direction, the current tends to flow around the grain centers, parallel to the laminations, and the field at the center is consequently feeble. By contrast, when the laminations are directed towards the grain centers, the current tends to be focused there.

One could similarly calculate threshold exponents for Schulgasser's three-dimensional symmetric material composed of spherical grains with radial laminations, [5] or for geometries with conelike features. Since the field can be highly concentrated (or feeble) near the grain center or near the apex of a cone, the threshold exponents will tend to be larger for these structures, as the field is effectively concentrated over a smaller proportion of space, compared with the corresponding two-dimensional structures.

To construct a laminate material with low threshold exponents, we need to consider hierarchical models since the field in any laminate material of finite rank is bounded and consequently has the trivial threshold exponents, $\gamma^+ = \infty$, $\gamma^- = -\infty$. The laminations need to be orientated so the maximum current is channeled into a small region, which suggests the hierarchical model of Fig. 8. When the current flow is directed parallel to the poor conducting laminations, and thus across the good conducting laminations, this material has the threshold exponents

$$\gamma^+ = 2/[1 - (\sigma_2/\sigma_1)^{1/2}], \ \gamma^- = -\infty \ , \tag{4.13}$$

and when the applied current is in the orthogonal direction we have

$$\gamma^+ = \infty \ , \ \gamma^- = 2/[(\sigma_1/\sigma_2)^{1/2} - 1]. \tag{4.14}$$

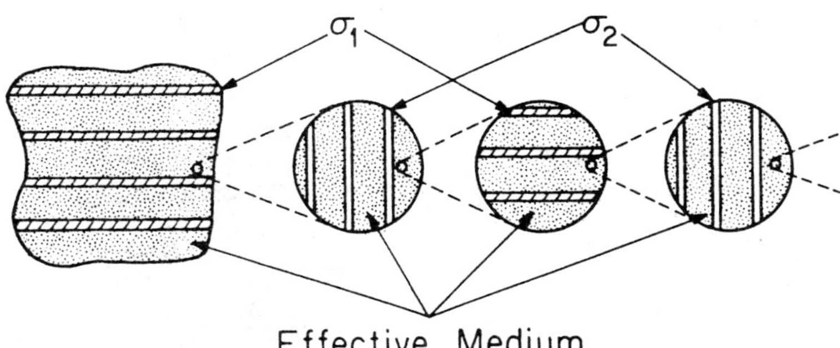

Fig. 8. A hierarchical laminate material with especially low threshold exponents. Equal infinitesimal proportions of the two components are introduced in
alternate order in the hierarchy. The slabs of component 1, which are located
at odd levels in the hierarchy, are placed orthogonal to the slabs of component
2, which are located at even levels in the hierarchy. Despite the anisotropic
construction this material has an isotropic conductivity $\sigma^* = (\sigma_1\sigma_2)^{1/2}$, given
exactly by the effective medium approximation [59],

The threshold exponents of the various materials we have considered are
compared in Fig 9. Note that the hierarchical laminate material has the lowest
threshold exponents. These few results, although selective, suggest that the
worst possible dielectric may, in fact, be a laminate material, possibly even
the one sketched in Fig. 8. Of course this is by no means proved. Many other
geometries, with a self-similar or fractal-like construction, should be examined
before any firm conclusions can be drawn.

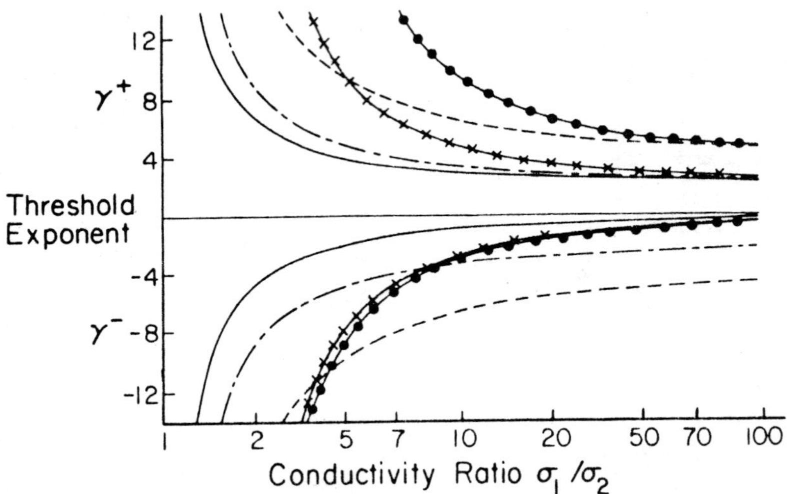

Fig. 9. Comparison of threshold exponents for the laminate of Fig. 8.
(——————), eqs. (4.13) and (4.18); an array of diamond shaped grains
(- - - - -), eq. (4.8); a checkerboard of the two components (——— - ———),
eq. (4.9); and Schulgasser's symmetric material, both in three dimensions
(⊖ ⊖ ⊖ ⊖) and in two dimensions (✕ ✕ ✕ ✕ ✕), eqs. (4.11) and (4.12).

Acknowledgements

I thank Dr. G.A. Baker, Jr., and Dr. J.G. Berryman for suggesting a connection between work on Pade approximants and bounds on conductivity functions and am grateful to J. Korringa for drawing my attention to Brugqeman's early work on laminates. I have also benefited from discussions with Professors B. Halperin, R.V. Kohn and G.A. Papanicolaou and with Dr. P.R. Graves-Morris and Prof. L. Tartar. The support of the California Institute of Technology, through a Weingart Fellowship, and the support of Chevron Laboratories, through a research grant, is greatly appreciated.

References

1.   K. Schulgasser, "Relationship between single-crystal and polycrystal electrical conductivity," J. Appl. Phys $\underline{47}$, 1880-1886 (1976).

2.   G.E. Backus, "Long-wave elastic anistropy produced by horizontal layering," J. Geophys. Res. $\underline{67}$, 4427-4440 (1962).

3.   L. Tartar, "Estimations fines des coefficients homogénéisés," in Pitman Research Notes in Mathematics, ed. P. Kree, 1985. See also the article in these proceedings.

4.   K. Schulgasser, "Bounds on the conductivity of statistically isotropic
     polycrystals," J. Phys. C 10, 407-417 (1977).

5.   K. Schulgasser, "Sphere assemblage model for polycrystals and symmetric
     materials," J. Appl. Phys. 54, 1380-1382 (1982).

6.   K.A. Lurie and A.V. Cherkaev, "Optimal structural design and relaxed
     controls," Opt. Control Appl. Meth. 4, 387-392 (1983).

7.   G.W. Milton, "Bounds on the complex permittivity of a two-component com-
     posite material," J. Appl. Phys. 52, 5286-5293 (1981).

8.   G.W. Milton, "Bounds on the transport and optical properties of a two-
     component composite material," J. Appl. Phys. 52, 5294-5304. (1981).

9.   K.A. Lurie and A.V. Cherkaev, "Accurate estimates of the conductivity of
     mixtures formed of two materials in a given proportion (two-dimensional
     problem)," Dokl. Adad. Nauk. 264, 1128-1130 (1982).

10. F. Murat and L. Tartar, "Calcul des variations et homogénéisation," to
     appear in Cours de l'Ecole d'Eté d'Analyse Numérique CEA-EDF-INRIA sur
     l'homogénéisation: Collection de la Direction des Etudes et Recherches
     d'Electricité de France, Eyrolles, Paris, (1984).

11. K.A. Lurie and A.V. Cherkaev, "Exact estimates of conductivity of a binary
     mixture of isotropic compounds," preprint.

12. Z. Hashin and S. Shtrikman, "A variational approach to the theory of effec-
     tive magnetic permeability of multiphase materials," J. Appl. Phys. 33,
     3125-3131 (1962).

13. Z. Hashin and S. Shtrikman, "A variational approach to the theory of the
     elastic behavior of multiphase materials," J. Mech. Phys. Solids, 11,
     127-140 (1963).

14. P.N. Sen, C. Scala, and M.H. Cohen, "A self-similar model for sedimentary
     rocks with application to the dielectric constant of fused glass beads,"
     Geophysics, 46, 781-795 (1981).

15. R.V. Kohn and G. Strang, "Optimal design and relaxation of variational
     problems," preprint.

16. G.A. Baker, Jr., "Best error bounds for Padé approximants to convergent
     series of Stieltjes," J. Math. Phys., 10, 814-820 (1969).

17. M.J. Beran, "Field fluctuations in a two-phase random medium," J. Math.
     Phys. 21, 2583-2585 (1980).

18. M.P. Bendsøe, "Optimization of Plates," thesis, Math. Inst. Tech. Univ.
     Denmark, Lyngby, Denmark.

19. K.A. Lurie and A.V. Cherkaev, "G-closure of a set of anisotropically con-
     ducting media in the two-dimensional case," Dokl. Akad. Nauk. 259, 328-331
     (1981).

20. K.A. Lurie and A.V. Cherkaev, "G-closure of a set of anistropically con-
     ducting media in the two-dimensional case," J. Opt. Th. Appl. 42, 283-304
     (1984).

21. A.M. Dykhne, "Conductivity of a two-dimensional two-phase system," Zh.
     Eksp. Teor, Fiz. 59, 110-115 (1970). [Soviet Phys. JETP 32, 63-65 (1971).]

22. K.S. Mendelson, "A theorem on the effective conductivity of a two-dimensional heterogeneous medium," J. Appl. Phys. $\underline{46}$, 4740-4741 (1975).

23. J.B. Keller, "A theorem on the conductivity of a composite medium," J. Math. Phys. $\underline{5}$, 548-549 (1964).

24. L. Tartar, "Compensated compactness and applications to partial differential equations," in Nonlinear Analysis and Mechanics, Research Notes in Mathematics $\underline{39}$, 136-212, Pitman (1979).

25. R.V. Kohn and G.W. Milton, "Bounds for anisotropic composites by variational principles," to be submitted; see also these proceedings.

26. G.W. Milton and K. Golden, "Thermal conduction in composites", in Thermal Conductivity 18, Plenum Press, New York (1985).

27. D.J. Bergman, "Rigorous bounds for the complex dielectric constant of a two-component composite," Annals of Physics $\underline{138}$, 78-114 (1982).

28. G.W. Milton, "Concerning bounds on the transport and mechanical properties of multicomponent composite materials," Appl. Phys. $\underline{A26}$, 125-130 (1981).

29. Z. Hashin and B.W. Rosen, "The elastic moduli of fiber-reinforced materials," J. Appl. Mech. Trans. ASME $\underline{31}$, 223-232 (1964).

30. K.H. Lo and R.M. Christensen, "Solutions for effective shear properties in three phase sphere and cylinder models," J. Mech. Phys. Solids $\underline{27}$, 315-330 (1979).

31. G.W. Milton and N. Phan-Thien, "New bounds on effective elastic moduli of two-component materials," Proc. R. Soc. Lond. $\underline{A380}$, 305-331 (1982).

32. A.N. Norris, "A differential scheme for the effective moduli of composites," to appear in Mechanics of Materials.

33. K.A. Lurie and A.V. Cherkaev, "The problem of formation of an optimal multicomponent composite," preprint.

34. D.J. Bergman, "The dielectric constant of a composite material - a problem in classical physics," Phys. Rep $\underline{C43}$, 377-407 (1978).

35. K. Golden and G. Papanicolaou, "Bounds for effective parameters of multicomponent media by analytic continuation", to appear in J.Stat. Phys.

36. K. Golden and G. Papanicolaou, "Bounds for effective parameters of heterogeneous media by analytic continuation," Comm. Math. Phys. $\underline{90}$, 473-491 (1983).

37. R.M. Forster, "A reactance theorem", Bell System Tech. J. $\underline{3}$, 259 (1924).

38. J.E. Storer, "Passive network synthesis," 15-37, McGraw Hill, New York (1957).

39. G.A. Baker, Jr., "Essentials of Padé approximants," Academic Press, New York (1975).

40. D.J. Bergman, "Exactly solvable microscopic geometries and rigorous bounds for the complex dielectric constant of a two-component composite material," Phys. Rev. Lett., $\underline{44}$, 1285-1287 (1980).

41. D.J. Bergman, "Bounds for the complex dielectric constant of a two-component composite material," Phys. Rev. B, $\underline{23}$ 3058-3065 (1980).

42. G.W. Milton, "Bounds on the complex dielectric constant of a composite material," Appl. Phys. Lett. 37, 300-303, (1980).

43. R.C. McPhedran, D.R. McKenzie and G.W. Milton, "Extraction of structural information from measured transport properties of composites," Appl. Phys. A 29, 19-27 (1982).

44. R.C. McPhedran and G.W. Milton, "Bounds and exact theories for the transport properties of inhomogeneous media," Appl. Phys. A26, 207-220 (1981).

45. B.U. Felderhof, "Bounds for the complex dielectric constant of a two-phase composite," Physica 126A, 430 (1984).

46. K. Golden, "Bounds for the complex permittivity of a multicomponent material," to appear in J. Mech. Phys. Solids.

47. M. Beran, "Use of a variational approach to determine bounds for the effective permittivity of a random medium," Nuovo Cimento 38, 771-782 (1965).

48. G.W. Milton and R.C. McPhedran, "A comparison of two methods for deriving bounds on the effective conductivity of composites," Lecture notes in Physics 154, 183-193, ed. by R. Burridge, S. Childress and G. Papanicolaou, Springer Verlag, New York, 1982.

49. S. Torquato and G. Stell, "Bounds on the effective thermal conductivity of a dispersion of fully penetrable spheres," to appear in Int. J. Eng. Sci.

50. S. Torquato and J.D. Beasley, "Effective properties of fibre-reinforced thermal conductivity of dispensions of fully penetrable cylinders," to appear in Int. J. Eng. Sci.

51. J.G. Berryman, "Variational bounds on elastic constants for the penetrable sphere model," J. Phys. D: Appl. Phys. 18, 585-597 (1985).

52. B.U. Felderhof, "Bounds for the effective dielectric constant of disordered two-phase materials," J. Phys. C15, 1731-1739 (1982).

53. A.K. Common, "Padé approximants and bounds to series of Stieltjes", J. Math. Phys. 9, 32-38 (1968).

54. Y. Kantor and D.J. Bergman, "Improved rigorous bounds on the effective elastic moduli of a composite material", J. Mech. Phys. Solids, 32, 41-62 (1984).

55. G.A. Baker, Jr., "The theory and application of the Padé approximant method", Advan. Theor. Phys. 1, 1-58 (1965).

56. M. Born and E. Wolf, "Principles of Optics", 708-718, Macmillan, New York (1964).

57. E.H. Kerner, "The electrial conductivity of composite media", Proc. Phys. Soc. B69, 805-806, (1956).

58. L.J. Walpole, "On the overall elastic moduli of composite materials", J. Mech. Phys. Solids, 247, (1964).

59. R. Landauer, "Electrical conductivity in inhomogeneous media", in Electrical Transport and Optical Properties of Inhomogeneous Media, 1-45, ed. J.C. Garland and D.B. Tanner, Amer. Inst. Phys., New York (1978).

60. J. Meixner, N.Y. Univ. Inst. Math. Sci. Research Rep. No. EM-72 (1955).

61. M. Soderberg and G. Grimvall, "Current distributions for a two-phase material with chequer-board geometry", J. Phys. C. $\underline{16}$, 1085-1088 (1983).

62. G.W. Milton, "A continued fraction representation and bounds for the conductivity of multicomponent anisotropic media", to be submitted.

63. P. Henrici and P. Pfluger, "Truncation error estimates for Stieltjes fractions", Num. math. $\underline{9}$, 120-138 (1966).

64. G.A. Baker, Jr., and P.R. Graves-Morris, Encyclopedia of mathematics and its applications, Vols. $\underline{13}$ and $\underline{14}$, ed. G.-C. Rota, Addison-Wesley, London (1981).

65. G.A. Francfort and F. Murat, "Homogenization and optimal bounds in linear elasticity," submitted to Arch. Rat. Mech. Anal.

66. D.A.G. Bruggeman, "Berechnung verschiedener physikalischer Konstanten, von Heterogenen Substanzen", Ann. Phys. $\underline{5}$, 636-664 (1935).

67. D.A.G. Bruggeman, "Elastizität Konstanten von Kristallaggregaten" Ph.D. Thesis, Utrecht, 19-38 (1930).

Appendix: Stieltjes Functions and Bounds

Here we discuss the close connection between transformations used by Bergman [27] and Golden [46] to derive bounds on the effective dielectric constant (or conductivity) and previous work of Baker [16] on bounds on Stieltjes functions: see also ref. 26. Given a composite with effective dielectric constant ε^*_{jj} in direction x_j , composed of two components with dielectric constants ε_1 , ε_2 , Bergman [34] and Golden and Papanicolaou [36] have established that the function

$$F(s) \equiv 1 - \varepsilon^*_{jj}/\varepsilon_2 \ , \quad s \equiv \varepsilon_2/(\varepsilon_2 - \varepsilon_1) \qquad (A.1)$$

has the integral representation

$$F(s) = \int_0^1 \frac{\mu(y)dy}{s-y} \ , \qquad (A.2)$$

where the measure $\mu(y)$ is real positive. Now, introducing the variable

$$z \equiv -1/s \ , \qquad (A.3)$$

it is clear that

$$h(z) \equiv -F(z)/z = \int_0^1 \frac{\mu(y)dy}{1 + yz} \qquad (A.4)$$

defines a Stieltjes function of z with a radius of convergence of at least 1.

The bounds of Baker [16] are based on the observation that if $h(z)$ is a Stieltjes function with radius of convergence R and series expansion

$$h(z) = h_0 + h_1 z + O(z^2) \tag{A.5}$$

then $h^{(1)}(z)$ defined implicitly by

$$h(z) = h_0/[1 + zh^{(1)}(z)] \tag{A.6}$$

is, in fact, also a Stieltjes function with the same radius of convergence, R. By comparison, Bergman [27] and Golden [46] find that if $F(s)$ has the asymptotic expansion

$$F(s) = a_1/s + a_2/s^2 + O(1/s^3) \tag{A.7}$$

then

$$F^{(1)}(s) \equiv 1/a_1 - 1/sF(s) \tag{A.8}$$

shares many of the important analytic properties of $F(s)$. Note that (A.6) can be expressed in the form

$$-zh^{(1)}(z)/h_0 = 1/h_0 - 1/h(z) \tag{A.9}$$

and may be regarded as a transformation mapping the Stieltjes function $h(z)$ into a new Stieltjes function, $h^{(1)}(z)/h_0$. The transformation (A.8) introduced by Bergman is thus equivalent to the transformation (A.9), with the identification

$$F^{(1)}(z) = -zh^{(1)}(z)/h_0 \, , \, h_0 = a_1 \, . \tag{A.10}$$

This transformation is important because any elementary bounds on the function $F^{(1)}(s)$ or $h^{(1)}(z)$ are mapped into more sophisticated bounds on $F(s)$ and $h(z)$ that incorporate additional information, namely the value of $h_0 (\text{or } a_0)$. A whole sequence of nested bounds can be generated by iterating this transformation in the manner described by Baker [16] and Golden [46], and suggested by Kantor and Bergman [54]: the nth order bounds incorporate all the coefficients after the original work on Stieltjes functions by Henrici and Pfluger [63], Common [53] and Baker [16].

It now appears, that the most natural way of deriving this hierarchy of bounds for the effective dielectric constant is to introduce a special transformation, similar to (2.2), that treats the two components on a symmetric basis [26]. The resulting bounds depend on a sequence of <u>normalization constants,</u> and <u>weights</u> that characterize the geometry of the composite. For a two-dimensional

WAVES IN BUBBLY LIQUIDS

R. Caflisch, M. Miksis*, G. Papanicolaou, L. Ting

Courant Institute, New York University

1. Introduction

Consider a gas-bubble liquid mixture with β the gas volume fraction p the pressure and ρ the density of the mixture. Let c_{eff} be the effective sound speed of the mixture and $\tau = \rho^{-1}$ the specific volume. We have that

$$(1.1) \qquad c_{eff}^2 = \frac{dp}{d\rho} = \frac{1}{d\rho/dp} = \frac{1}{d\tau^{-1}/dp} = \frac{1}{\kappa\rho}$$

where the compressibility κ is defined by

$$(1.2) \qquad \kappa = -\frac{1}{\tau}\frac{d\tau}{dp}$$

i.e. the change of volume with respect to pressure. Now let us assume that density and compressibility of the mixture are simply the averages over the two component values

$$(1.3) \qquad \rho = \rho_g\beta + \rho_\ell(1-\beta) \ , \ \kappa = \kappa_g\beta + (1-\beta)\kappa_\ell \ ,$$

where subscripts denote liquid or gas. The density of the gas is typically 1000 times smaller than that of the liquid while the compressibility of the liquid is negligible. Combining (1.1) and (1.3) with this simplification gives the formula

$$(1.4) \qquad c_{eff}^2 = \frac{\kappa_g^{-1}}{\rho_\ell\beta(1-\beta)}$$

If now $p = \text{const.} \ \rho^\gamma$ for the gas with γ the ratio of specific heats, we have $\kappa_g^{-1} = \gamma p$ and hence

$$(1.5) \qquad c_{eff}^2 = \frac{\gamma p}{\rho_\ell\beta(1-\beta)} \qquad .$$

---

* Dept. of Mathematics, Duke University.

This is the well known formula [1] for the effective sound speed that was derived early in this century. It is a simple but striking result because at pressure $p = 10^6$ dynes/cm^2 , $\rho_\ell = 1$ gm/cm^3 , $\gamma = 1$ and $\beta = 10^{-2}$ (1% gas bubble volume fraction) we get $c_{eff} \sim 100$ m/sec . Thus the sound speed in the mixture is about 14 times lower than that in the liquid and more than 3 times lower than that in the gas. A small amount of gas in the liquid changes its effective properties in a remarkable way. The reasons for this are clear by looking at (1.1) and (1.3). The bubbly liquid is it at once massive, because it is mostly liquid, and compressible because it gets its compressibility from the small amount of gas in it. Gas with liquid drops in it behaves in a more usual way as a mixture, regarding sound propagation.

The above discussion indicates that an interesting theory should be available for describing sound propagation in bubbly liquids. There is considerable literature on this subject especially since the nineteen forties (cf. for example [2],[3] and the review [1]).

We have been interested in the analytical understanding of the following four main questions.

(1) Can one derive a set of effective equations for sound propagation in a bubbly liquid?

(2) How does nonlinearity, i.e. large pressure variation, affect the derivation of the effective equations?

(3) Can effective equations be obtained at larger gas-bubble volume fractions?

(4) What are the analytical properties of the, usually nonlinear, effective equations?

We have made some progress in these questions [4,5,6,7,8]. Questions (1) and (3) are well understood in the linear regime [5] even though this does not seem to be very well known. In the nonlinear regime only low volume fraction analysis has been given [4] which recovers a set of effective equations written down without a derivation from first principles by van Wijngaarden [1]. Mathematical analysis of the nonlinear effective equations has been given in one space dimension [6]. Larger volume fraction effects and nonlinear effects have been analyzed by Rubinstein [8] for a periodic configuration of bubbles.

2. The Microscopic Problem.

Supppose that N spherical gas bubbles are located at $x_1, x_2, \ldots, x_N$ with radii $R_1(t), R_2(t), \ldots, R_N(t)$. The equations of motion in the liquid $|x-x_j| > R_j$, $j=1, \ldots, N$ are

$$(2.1) \qquad \frac{1}{\rho_\ell c_\ell^2} \, (p_t + u \cdot \nabla p) + \nabla \cdot u = 0$$

$$(2.2) \qquad \rho_\ell (u_t + u \cdot \nabla u) + \nabla p = 0$$

with $\nabla \times u = 0$ assumed throughout. The boundary conditions are

$$(2.3) \qquad \frac{\partial R_j}{\partial t} = u \cdot \hat{n} \qquad (\hat{n} \text{ is the unit outward normal})$$

$$(2.4) \qquad p = p_g \qquad \text{on } |x-x_j| = R_j(t) , \ j = 1, \ldots, N$$

with

$$(2.5) \qquad p_g = \kappa \left(\frac{M_j}{\frac{4}{3} \pi R_j^3} \right)^\gamma$$

Here ρ_ℓ , c_ℓ , M_j , κ and γ are constants: the liquid density, liquid sound speed, mass of jth gas bubble, constant in equation of state and ratio of specific heats, respectively. Several simplifications, discussed more fully in [4] , have been made here which are not needed in describing the passage to a continuum liquid.

We assume that the bubble centers $\{x_1^N, \ldots, x_N^N\}$ are a fixed set of points, the Nth configuration. N will be taken to infinity and we will assume that

$$\frac{1}{N} \theta^N(A) = \frac{1}{N} \{\# \text{ of points } x_j^N \text{ in set } A\}$$

satisfies

$$(2.6) \qquad \frac{1}{N} \theta^N(A) \to \int_A \theta(x) \, dx$$

where $\theta(x)$ is the continuum bubble center density. Note that the bubble centers do not move. This is appropriate for wave propagation phenomena.

To pass to a continuum limit in (2.1) - (2.5) a careful scaling is needed. Suppose V is the volume of the region occupied by the bubbles and λ is a typical wavelength of a propagating disturbance. Let

$$(2.7) \qquad \varepsilon = \frac{1}{\lambda} \left(\frac{V}{N} \right)^{1/3} = \frac{\text{interbubble distance}}{\text{wave length}}$$

$$(2.8) \qquad \beta = R_0/\lambda = \frac{\text{typical bubble radius}}{\text{wave length}}$$

$$(2.9) \qquad \beta = \frac{1}{V} \frac{4}{3} \pi R_0^3 N = \frac{4}{3} \pi \left(\frac{\delta}{\varepsilon} \right)^3 = \text{volume fraction}$$

$$(2.10) \qquad \chi = \frac{N}{V} \lambda^3 \delta = \frac{\delta}{\varepsilon^3}$$

The two dimensionless parameters ε and δ control the analysis. Both are going to zero in the continuum limit but we can have it so that β stays fixed or χ stays fixed. The one case is finite or "large" volume fraction while the other corresponds to small volume fraction.

The limit $\delta \to 0$, $\varepsilon \to 0$, β fixed is analyzed in [5] in the _small_ amplitude regime and in [8] more generally but for periodic configurations. The limit $\delta \to 0$, $\varepsilon \to 0$, χ fixed is analyzed in [4] and produces the effective equations of van Wijugaarden [1]. It is surprising that it was not known previously that the equations in [1] arise in this manner i.e. by keeping the number χ fixed in the limit.

We describe now these limiting equations. We may do so in terms of a velocity potential ϕ with $u = \nabla\phi$, because we assume $\nabla \times u = 0$. The velocity potential in the limit $N \to \infty$ $(\varepsilon \to 0)$, $\delta \to 0$ and χ fixed tends to the solution of

$$(2.11) \qquad C^{-1} \phi_{tt} - \Delta\Phi + \left(\frac{4}{3} \pi R^3 \theta X \right)_t = 0$$

$$(2.12) \qquad \nabla\phi(0,x) = u_0(x) \qquad \text{initial velocity.}$$

$$(2.13) \qquad \phi_t(0,x) = -\varsigma\, p_0(x) \qquad \text{initial pressure}$$

with

$$(2.14) \qquad RR_{tt} + \frac{3}{2} R_t^2 = \zeta(F(R) + \frac{1}{\zeta} \overline{\phi}_t)$$

$$(2.15) \qquad F(R) = (\frac{M}{R^3})^\gamma$$

We have written the limit equations in dimensionless form so C and ζ are dimensionless numbers (which are given in [4]). We want to point out here that the limit equations are a wave-equation coupled to a nonlinear ODE in what appears to be a rather nonobvious way, in view of where one starts (2.1) - (2.5). Actually it is fairly clear how the equations emerge once one looks at (2.1) - (2.5) more closely [4]. The key idea is to use Foldy's method [2], suitably modified to account for nonlinear effects. Foldy's method says, roughly, that at small volume fraction each bubble sees only the average or macroscopic field driving it and not the complicated local fields of all the other bubbles.

3. Mathematical questions

Making precise mathematically Foldy's aproximation is quite complicated. For linear problems of wave propagation and, primarily, for diffusion it has been analysed extensively [9,10,11]. For nonlinear problems like the one described in section 2 this looks very difficult. In [4] we derived (2.11) - (2.15) systematically and carefully but not in a rigorous way.

To get experience with the mathematics of the nonlinear Foldy approximation we made up [12] the following problem that is perhaps of independent interest. Given a configuration $(x_1^N,\ldots,x_N^N)$ of points in R^3, solve for $u^N(t,x)$ and $R_j^N(t)$, $j = 1,\ldots,N$ in

$$(3.1) \qquad u_t^N = \Delta u^N \qquad \text{for } |x-x_j^N| > \delta R_j^N(t)$$

$$u^N(0,x) = f_0(x) \qquad j = 1,\ldots,N \; ,$$

$$(3.2) \qquad u^N = 0$$

$$- \frac{1}{\delta} \frac{\partial}{\partial t} R_j^N(t) = \sigma \int_{|x-x_j^N| = \delta R_j^N(t)} \frac{\partial u}{\partial n} \quad \text{on} \quad |x - x_j^N| = \delta R_j^N(t)$$

$$R_j^N(0) = R_0(x_j^N) \, .$$

Here δ is a parameter and $N \to \infty$, $\delta \to 0$ so that $N\delta = 1$. The configuration $(x_1^N,\ldots,x_N^N)$ satisfy (2.6) and another condition spelled out in [11,12].

Now we can prove a theorem [12] that $u^N(t,x)$ converges uniformly, outside of a set that has zero volume in the limit $N \to \infty$, $\delta \to 0$, $\delta N = 1$, to the solution of the equation

$$(3.3) \qquad \overline{u}_t = \Delta \overline{u} - 4\pi R \overline{u} \, , \quad \overline{u}(0,x) = f_0(x)$$

$$(3.4) \qquad - R\frac{\partial R}{\partial t} = \sigma \overline{u} \, , \quad R(0,x) = R_0(x)$$

Note that (3.4) is an ODE that can be solved and plugged into (3.3) to give

$$(3.5) \qquad \overline{u}_t = \Delta \overline{u} - 4\pi \left[R_0(x) - 2\sigma \int_0^t \overline{u}(s,x)ds \right]^{1/2} \overline{u}$$

$$\overline{u}(0,x) = f_0(x).$$

A physical interpretation of (3.1) - (3.2) is this. The function $u(t,x)$ is the temperature of water which has spheres of ice imbedded in it. The ice forces the temperature of the water to be zero on its surface. The ice also melts because it absorbs heat. This is the content of (3.2). In the continuum limit all the phenomena are contained in (3.5) at a macroscopic and much simpler level. Complete details of this problem are in [12].

References

[1] L. van Wijngaarden, One-dimensional flow of liquids containing small gas bubbles, Ann. Rev. Fluid. Mech. (1972) 369-394.

[2] E.L. Carstensen and L. Foldy, Propagation of sound through a liquid containing bubbles, JASA 19 (1947) 481-501.

[3] G.B. Wallis, One-dimensional two-phase flow, McGraw-Hill, 1969.

[4] R.E. Caflisch, M. Miksis, G. Papanicolaou and L. Ting, Effective equations for wave propagation in bubbly liquids, J. Fluid Mechanics, to appear.

[5] R.E. Caflisch, M. Miksis, G. Papanicolaou and L. Ting, Wave propagation in bubbly liquids at small and finite volume fraction, J. Fluid Mechanics.

[6] R.E. Caflisch, Global existence for a nonlinear theory of bubbly liquids, Comm. Pure Appl. Math, to appear.

[7] M. Miksis and L. Ting, Nonlinear radial oscillations of a gas bubble including thermal effects, JASA, 1984.

[8] J. Rubinstein, Bubble interaction effects on waves in bubbly liquids, JASA, 1985.

[9] E.I. Hruslov and V.A. Marchenco, Boundary value problems in regions with fine-grained boundaries, Naukova Dumka, Kiev, 1974.

[10] S. Ozawa, On an elaboration of M. Kac's Theorem concerning eigenvalues of the Laplacian in a region with randomly distributed small obstacles, Comm. Math. Phys., 1983.

[11] G. Papanicolaou and S.R.S. Varadhan, Diffusion in regions with many small holes, in Stochastic Differential Systems, B. Grigelions editor, Springer Lecture notes in Control and Info # 25 (1980) 190-206.

[12] G. Papanicolaou and J. Rubinstein, to appear.

SOME EXAMPLES OF CRINKLES

A.C. Pipkin
Division of Applied Mathematics
Brown University
Providence, RI   02912

1. Introduction

Many physical problems can be phrased as the problem of minimizing some energy functional $E[f]$ over a given class of admissible functions f. It can happen that there is a minimizing sequence f_n that approaches a limit $\bar{f}$ but $\bar{f}$ does not minimize E, either because $\bar{f}$ is not in the admissible class or because E is not lower semicontinuous. In the examples that I discuss here, this happens because the derivatives f'_n are highly discontinuous and do not approach $\bar{f}'$ in the limit. I call such sequences _crinkles_, and call the limiting function $\bar{f}$ the _carrier_ of the crinkle. Young [1] has written a book on the subject; he calls such sequences _generalized curves_. In control theory the same sort of thing is also called a _chattering state_.

In some cases all of the physical information that is wanted is embodied in the carrier $\bar{f}$. It then makes sense to try to rephrase the problem in terms of $\bar{f}$ alone. This may involve widening the class of admissible functions, since functions admissible as carriers may not all be admissible by the original standard. In such cases it is necessary to determine the energy E that is to be attributed to a previously inadmissible carrier. I call this _completion._ On the other hand, we may wish to determine the least energy that can be attributed to a carrier $\bar{f}$ by computing the energies in sequences approaching $\bar{f}$. This is known as _relaxation._ In either case we find a new functional $E^*[\bar{f}]$, defined in terms of $E[f]$ by a process that amounts to averaging over small regions within which f' may oscillate rapidly and discontinuously. This homogenization process is explained in some detail in Section 2, where we consider an example of relaxation. An example of completion is discussed in Section 3.

In Sections 4 and 5 we show how crinkles arise in a natural way in a theory of sheets formed from inextensible fibers [2]. In this application the crinkle has the interpretation of a continuously distributed family of wrinkles in the sheet.

Some matters involving crinkles in finite elasticity theory, from the work of Morrey [3] and Dacorogna [4], are discussed in Section 6.

2. An Example of Relaxation.

Young [1] has used an example that is nearly the same as the following one. Consider the functional

$$E[f] = \int_0^1 [(f'^2 - 1)^2 + f^2]dx,$$

which is to be minimized subject to the constraint $f(0) = f(1) = 0$. We can make E equal to zero by taking $f' = \pm 1$ and $f = 0$, but these conditions are contradictory unless we can some how divorce f from its derivative. The conditions $f' = \pm 1$ and $f = 0$ are trying to tell us that we can form a minimizing sequence f_n by taking f_n to be a saw-tooth function with slopes $f_n' = \pm 1$ and taking the intervals with $f_n' = 1$ to alternate very rapidly with those on which $f_n' = -1$, so that f_n never has a chance to be much different from $\overline{f} = 0$.

If the problem had arisen in a physical context, it is possible that the lack of a minimizer might make physical sense, in which case there would be nothing further to say. Another possibility is that the energy _ought_ to have a minimzer, and that the energy functional should be modified by taking into account physical effects that were thought to be negligible. But I want to consider a third possibility, that $\overline{f} = 0$ is really an acceptable answer, but $E[f]$ does not express this fact in the most convenient form.

Consider a more general problem in which $f(1) = c \neq 0$, so that $\overline{f} = 0$ is not the right answer. For any given carrier $\overline{f}(x)$, define a sequence approaching $\overline{f}$ in the following way. Partition the interval $[0,1]$ into subintervals that are labelled $(+)$ and $(-)$ altenately. Define a sequence of such partitionings, in which the number of subintervals approaches infinity while their lengths approach zero. Let $I_n(x)$ be the indicator function for the $(+)$ intervals in the n-th partitioning, equal to 1 when x is in a (+) interval and 0 when it is in a (-) interval. We require that the sequence of partitionings have a certain regularity to it, such that

$$\lim \int_a^b I_n(x)dx = \int_a^b \lambda(x)dx$$

for any (a,b). Thus $\lambda(x)$ represents the limiting fractional density of (+) intervals near x. We note that

$$0 < \lambda(x) < 1.$$

It follows that for any continuous $f(x)$,

$$\lim_n \int_a^b I_n(x)f(x)dx = \int_a^b \lambda(x)f(x)dx.$$

Now let $f'_+(x)$ and $f'_-(x)$ be two arbitrary continuous functions. Let f'_n be f'_+ on (+) intervals and f'_- on (-) intervals of the n-th partitioning. Then

$$f_n(x) = \int_0^x [I_n(y)f'_+(y) + (1 - I_n)f'_-(y)]dy.$$

In the limit, f_n approaches $\overline{f}$, where

$$\overline{f}(x) = \int_0^x [\lambda(y)f'_+(y) + (1 - \lambda)f'_-(y)]dy.$$

Then $\overline{f}'$ is

$$\overline{f}'(x) = \lambda(x)f'_+(x) + (1 - \lambda(x))f'_-(x), \tag{C}$$

an average of f'_+ and f'_-.

We similarly evaluate the limit of $E[f_n]$, and find that it is

$$\int [\lambda(x)(f'^2_+ - 1)^2 + (1 - \lambda)(f'^2_- - 1)^2 + \overline{f}^2]dx.$$

This accomplishes the separation of f from its derivative that we set out to do.

We now minimize over the extra degrees of freedom embodied in λ, f'_+ and f'_-. The part of the integrand involving these quantities has the form

$$\lambda W(f'_+) + (1 - \lambda)W(f'_-),$$

with a specific form of W , and we want to minimize this subject to the constraint (C). This can be done pointwise, and the result is a function of $\overline{f}'$ alone:

$$W_c(\overline{f}') = \min[\lambda W(f'_+) + (1 - \lambda)W(f'_-)].$$

Specifically, W_c is the greatest convex lower bound for the function W . In the present case it is

$$W_c(\overline{f}') = \begin{cases} (\overline{f}'^2 - 1) & \text{if } \overline{f}'^2 > 1, \\ \\ 0 & \text{if } \overline{f}'^2 < 1. \end{cases}$$

After this pointwise relaxation to lowest energy, the energy associated with $\overline{f}$ is

$$E^*[\overline{f}] = \int_0^1 [W_c(\overline{f}') + \overline{f}^2]dx.$$

The relaxed functional E^* agrees with the original one wherever there is no advantage in using a crinkle. If the original functional has a minimizer, the same function also minimizes E^* , but unlike E , E^* always has a minimizer.

3. An Example of Completion

The following example illustrates a different reason for using crinkles. Without yet specifying the form of $E[f]$, suppose that the constraints on admissible functions are $f(0) = 0$, $f(1) = c$, and $f'(x) = \pm 1$. That is, we admit only saw-tooth functions. We take $|c| < 1$, since otherwise no f is admissible.

We introduce a crinkle with $f'_+ = 1$ and $f'_- = -1$. The averaging condition for the carrier is

$$\overline{f}' = \lambda(+1) + (1 - \lambda)(-1),$$

and this determines λ in terms of $\overline{f}'$,

186

$$\lambda = (\overline{f}' + 1)/2.$$

Then the requirement that $0 < \lambda < 1$ puts a constraint on $\overline{f}'$, but not such a strict constraint as $f' = \pm 1$:

$$|\overline{f}'| < 1.$$

Here it is the constraint that is relaxed, not the energy functional.

Now suppose that the energy has the form

$$E[f] = \int_0^1 [W(f',x) + f^2(x)]dx.$$

W is defined by

$$W(1,x) = (1 - x)^2, \quad W(-1,x) = (1 + x)^2.$$

In the limit of a crinkle, the integral of W becomes the integral of

$$\lambda W(1,x) + (1 - \lambda)W(-1,x).$$

Expressing λ in terms of $\overline{f}'$ in the required way, we find that this is

$$1 + x^2 - 2x\overline{f}' .$$

Then integrating the term involving $\overline{f}'$ by parts, we find that the limiting value of the energy is

$$E^*[\overline{f}] = -2\overline{f}(1) + 1/3 + \int_0^1 (\overline{f} + 1)^2 dx.$$

In the present case we used a crinkle to find out what energy should be attributed to a function $\overline{f}$ for which the energy was not originally defined. The completion process introduces extra degrees of freedom by allowing functions $\overline{f}$ that were not originally admissible. In arriving at the functional E^* we have not yet minimized over these new degrees of freedom, so the relaxation part of the process has not been done yet.

4. Inextensible Networks

In the theory of networks formed from inextensible fibers [5,6], in some problems the energy has no kinematically admissible minimizer, but a convergent minimizing sequence exists, whose limit is not admissible. We demonstrate this by a specific example in the present section. The completion process is discussed further in the following section.

The body to be considered is a sheet that originally occupies a region B of the x,y plane. The material lines x = constant and y = constant are called fibers. In a deformation, the particle initially at (x,y) goes to $\underline{R}(x,y)$ in three-dimensional space. The vectors $\underline{A} = \underline{R}_x$ and $\underline{B} = \underline{R}_y$ are tangential to the deformed fibers, and the magnitudes of these vectors represent the local ratios of stretched to unstretched lengths. For inextensible networks there is no lengthening or shortening of fibers, and no deformations are admissible except those that satisfy the constraints

$$\underline{A} \cdot \underline{A} = \underline{B} \cdot \underline{B} = 1.$$

Let C_p and C_t be complementary, non-empty parts of the boundary of B. On C_p we specify $R(x,y) = \underline{R}_0(x,y)$. Admissible deformations are those that satisfy this boundary condition and the inextensibility conditions.

On C_t there is a force per unit initial length $\underline{T}$, applied as a dead load. For pure networks there is no strain energy, so the total energy of a deformation is the energy of the loads,

$$E[\underline{R}] = -\int_{C_t} \underline{R} \cdot \underline{T} ds,$$

where s is arc length along C_t. We seek to minimize E over admissible deformations.

Consider the following specific problem. B is the rectangular region $0 < x < L$, $0 < y < H$. The sheet is fixed along the edges $x = 0$ and $y = 0$, so that $\underline{R} = x\underline{i} + y\underline{j}$ there. On $y = H$ there is no traction. On $x = L$ the traction is

$$\underline{T}(y) = \underline{u}(\theta(y))T_0, \quad \underline{u}(\theta) = \underline{i} \cos \theta + \underline{j} \sin \theta,$$

where $T_0 > 0$ and the direction of the load is

$$\theta(y) = -\theta_0 y/L,$$

so that it slopes down more steeply as y increases.

The work done by the boundary load is maximized if each fiber y = constant swings around into the direction of the load on its end, so that $\underline{A} = \underline{u}(\theta(y))$. With the boundary condition on $x = 0$, this gives the deformation

$$\underline{r} = y\underline{j} + x\underline{u}(\theta(y)).$$

However, the derivatives of $\underline{r}$ are

$$\underline{a} = \underline{r}_x = \underline{u} \quad \text{and} \quad \underline{b} = \underline{r}_y = \underline{j} - (\theta_0 x/L)\underline{u}',$$

and $\underline{b}$ is not a unit vector so $\underline{r}$ is not an admissible deformation.

But it is easy to define a crinkle that approaches $\underline{r}$ as its carrier, and the crinkle is then a minimizing sequence. Divide the sheet into n horizontal strips of height H/n, so that the top edge of strip number k is at $y_k = kH/n$. Deform each strip in two steps. First, shear it so that $\underline{B} = \underline{j}$ and $\underline{A} = \underline{u}(\theta_k)$, i.e. the top edge goes into the direction of the boundary load at its end. Then fold the sheared strip so that the bottom edge coincides with the top edge of the next strip down. There is a second fold at the two edges that have been brought together, and so $2n$ folds altogether. This defines the admissible deformation $\underline{R}_n$. The sequence of deformations $\underline{R}_n$ approaches $\underline{r}$ pointwise. The derivatives $\underline{A}_n$, although discontinuous across the folds, approach $\underline{a} = \underline{r}_x$ in the limit, but the sequence $\underline{B}_n$ does not approach $\underline{b}$, and in fact $\underline{B}_n$ can be regarded as discontinuous everywhere in the limit.

The relaxed constraint conditions to be satisfied by carriers of crinkles are easy to derive, and easier still to guess. From the inextensibility conditions it follows that two particles (x_1,y_1) and (x_2,y_2) can never be separated further than the shortest length of fiber connecting them. With restriction to particles belonging to an arbitrary convex subset of B, this means that

$$|\underline{R}(x_1,y_1) - \underline{R}(x_2,y_2)| \leq |x_1 - x_2| + |y_1 - y_2|.$$

Now, if $\underline{R}_n$ is a convergent sequence of functions that all satisfy this Lipschitz condition, then the limit $\underline{r}$ satisifies it too. But the inequality does not guarantee that the derivatives $\underline{a} = \underline{r}_x$ and $\underline{b} = \underline{r}_y$ are unit vectors, but only that

$$\underline{a} \cdot \underline{a} \leq 1 \quad \text{and} \quad \underline{b} \cdot \underline{b} \leq 1.$$

No change in the form of the energy functional is required in order to define the energy of a carrier, because the functional does not involve the derivatives of the deformation. Because the relaxed constraint condition is a Lipschitz condition, every minimizing sequence has a subsequence that converges to an admissible deformation, and the energy at the limiting deformation is the minimum energy since the energy functional is continuous. So, relaxing the constraint is all that is necessary to guarantee existence of solutions. The resulting theory was originally put forward in its own right [7,8] rather than by the completion process discussed here.

5. Networks that Resist Distortion [2].

For a network of inextensible fibers like that described in Section 4, the inner product $\underline{A} \cdot \underline{B}$ is the sine of the angle of shear, and it measures the local distortion of the network. Let us suppose that there is a strain energy $W(\underline{A} \cdot \underline{B})$ associated with the distortion. Then the energy functional is

$$E = \int_B W(\underline{A} \cdot \underline{B})dxdy - \int_{C_t} \underline{R} \cdot \underline{T}ds.$$

We need to evaluate the strain energy term for the limit of a crinkle formed from inextensible deformations. The deformations $\underline{R}_n$ will represent folded states of the sheet, with the folds becoming more numerous and closer together as n increases. The folds divide the sheet into $(+)$ and $(-)$ strips. The derivatives $\underline{A}_n$ and $\underline{B}_n$ approach two limits each, in effect, $\underline{A}_+,\underline{B}_+$ and $\underline{A}_-,\underline{B}_-$. The derivatives $\underline{a}$ and $\underline{b}$ of the limit function $\underline{r}$ are averages of these limiting derivatives:

$$\underline{a} = \lambda \underline{A}_+ + (1 - \lambda)\underline{A}_-, \quad \underline{b} = \lambda \underline{B}_+ + (1 - \lambda)\underline{B}_-.$$

Now, $\underline{A}_n$ and $\underline{B}_n$ must satisfy certain continuity conditions at the folds, where they are discontinuous, sufficient to ensure that $\underline{R}_n$ is continuous across the fold. These are that the tangential components must be continuous and the magnitudes of the normal components must be continuous. If $\underline{t}$ is a unit vector tangential to the fold and $\underline{n}_+$ and $\underline{n}_-$ are unit vectors perpendicular to $\underline{t}$ and tangential to the sheet on the two sides of the fold, then the limits of $\underline{A}_n$ and $\underline{B}_n$ have the forms

$$\underline{A}_\pm = A_t \underline{t} + A_n \underline{n}_\pm, \quad \underline{B}_\pm = B_t \underline{t} + B_n \underline{n}_\pm.$$

It follows that the amount of shear is the same on both sides of the fold:

$$\underline{A}_+ \cdot \underline{B}_+ = \underline{A}_- \cdot \underline{B}_- = \underline{A} \cdot \underline{B} \quad (\text{say}).$$

If we wish to regard a given function $\underline{r}$ as the carrier of a crinkle, it turns out that the directions $\underline{t}$ of the fold lines are almost uniquely determined by $\underline{r}$. Let $y = f(x,c)$ be the fold loci in the original configuration. Then $\underline{t}$ is defined by $m\underline{t} = \underline{a} + f_x \underline{b}$, where m is the magnitude of the right-hand side. By using the continuity conditions, the averaging conditions, and the facts that $\underline{A}_\pm$ and $\underline{B}_\pm$ are unit vectors, it is possible to determine f_x up to a choice of sign, and so determine two possible choices for $\underline{t}$. With either of these choices, A_t, A_n, B_t and B_n are determined uniquely in terms of $\underline{a}$ and $\underline{b}$, so it is possible to evaluate $\underline{A} \cdot \underline{B}$ in terms of $\underline{a}$ and $\underline{b}$:

$$\underline{A} \cdot \underline{B} = \underline{a} \cdot \underline{b} \pm [(1 - \underline{a} \cdot \underline{a})(1 - \underline{b} \cdot \underline{b})]^{1/2}.$$

The choice of sign here comes from the two possible ways to represent $\underline{r}$ as the limit of a crinkle.

The crinkle is used in this example as a completion process, to determine the value of E that should be associated with functions that were not initially admissible. But a slight touch of relaxation is possible, because of the undetermined sign in $\underline{A} \cdot \underline{B}$. Assuming that W is an increasing function of $|\underline{A} \cdot \underline{B}|$, we minimize over the remaining degree of freedom, the $(\pm)$ sign, by choosing the sign that makes $|\underline{A} \cdot \underline{B}|$ smaller:

$$\underline{A} \cdot \underline{B} = \underline{a} \cdot \underline{b} - (\text{sgn } \underline{a} \cdot \underline{b})[(1 - \underline{a} \cdot \underline{a})(1 - \underline{b} \cdot \underline{b})]^{1/2}.$$

This is the expression for $\underline{A} \cdot \underline{B}$ that we use in $W(\underline{A} \cdot \underline{B})$ in order to define the extended energy functional $E^*[\underline{r}]$.

It would be nice to be able to say that E^* now always has a minimizer. Indeed it has a greatest lower bound E_0, presumably the same as that for E, and a minimizing sequence $\underline{r}_n$ that converges to a deformation $\underline{r}_0$ (say) that now is itself admissible, but we have not proved that $E^*[\underline{r}_0] = E_0$. This is true if E^* is lower semicontinuous, but we have not proved that it is.

This quandary is associated with the idea that a minimizing sequence $\underline{R}_n$ for the original functional E may involve functions with more complicated structure than those we have used in describing a crinkle. The kinds of crinkles that we have used are one-dimensional in the sense that we can number the places of discontinuity with one parameter, for example a one-parameter family of curves in the present example. It is conceivable that minimizing sequences are messier than that.

6. Finite Elasticity

In finite elasticity theory we deal with the following kind of minimization problem. Let $\underline{x}$ be the initial position of a particle in a body B. In a deformation the particle moves to the place $\underline{y}(\underline{x})$. The derivative $\underline{F}$ defined by $d\underline{y} = \underline{F} \, d\underline{x}$ is the deformation gradient. The strain energy density W is some specified function of $\underline{F}$, bounded below for all $\underline{F}$ and large when $\underline{F}$ is large. The total strain energy is

$$E = \int_B W(\underline{F}) dV.$$

We seek the deformation that minimizes this, subject to the constraint that $\underline{y}$

has specified values $\underline{y}_0(\underline{x})$, say, on the surface S of B.

Consider the more specific class of problems in which the boundary values have the form $\underline{y}_0 = \underline{F}\underline{x}$ with $\underline{F}$ constant. In that case the uniform deformation $\underline{y} = \underline{F}\underline{x}$ (in B) is admissible, and it is an equilibrium state, i.e. E is stationary at this deformation. The uniform deformation is <u>stable</u> if it minimizes E, and unstable otherwise.

For a problem of this kind, any admissible deformation can be written as $\underline{y} = \underline{F}\underline{x} + \underline{u}(\underline{x})$, with $\underline{u} = \underline{0}$ on S. Then the deformation gradient is $\underline{F} + (\nabla\underline{u})^t$. Let us call $\underline{F}$ the <u>apparent</u> deformation gradient. The <u>apparent energy density</u> $W^*(\underline{F},B)$ is defined by

$$W^*(\underline{F},B)V(B) = \inf \int_B W(\underline{F} + (\nabla\underline{u})^t)dV,$$

where the infimum is over all sufficiently smooth functions $\underline{u}$ that satisfy $\underline{u} = \underline{0}$ on S. $V(B)$ is the volume of the body.

Because W does not depend explicitly on $\underline{x}$ or $\underline{y}(\underline{x})$, W^* is invariant under translations of the body, and with the volume $V(B)$ taken into account as indicated, W^* is also invariant under scale changes. It depends on the shape of the body and its orientation, at most.

If there is some (most stable) body shape B_s for which the uniform deformation is stable, i.e. $W^*(\underline{F},B_s) = W(\underline{F})$, then we say that the <u>material</u> is stable at $\underline{F}$. Now, because of the scale and translation invariance just mentioned, any body B can be regarded as a subset of B_s, and it then follows that $W^*(\underline{F},B) = W(\underline{F})$. For otherwise, the energy-minimizing deformation in B plus the deformation $\underline{u} = \underline{0}$ in $B_s - B$ would give an energy for B_s lower than $W(\underline{F})V(B_s)$. Thus the stability property is independent of the body shape, and that is why we say that the <u>material</u> is stable at $\underline{F}$. As a corollary, if $W^*(\underline{F},B_u) < W(\underline{F})$ for some (most unstable) body shape B_u, then the same is true for all bodies, and we say that the <u>material</u> is unstable at $\underline{F}$.

By using more elaborate embedding arguments involving crinkles, Morrey [3] proved the remarkable result that whether or not the material is stable at $\underline{F}$, the apparent energy density is independent of the body shape:

$$W^*(\underline{F},B) = W^*(\underline{F}).$$

Thus $W^*(\underline{F})$ is a material property, determined by $W(\underline{F})$ alone without reference to any particular structure. Morrey did not find this interesting enough to label it as a separate lemma, but it is buried in his proof of something else.

The main idea in Morrey's proofs is that if a certain average energy density can be achieved in one body, it can also be achieved in any other body, and moreover this can be done in the limit of a crinkle that is approaching the uniform state. In one dimension this is very easy to show. Let the energy be

$$E[f] = \int_0^1 W(F \div u'(x))dx,$$

where $f = Fx + u(x)$ and the admissibility condition on u is $u(0) = u(1) = 0$. Given any admissible $u_1(x)$, define a sequence u_n in the following way: On $0 < x < 1/n$, let $u_n'(x) = u_1'(nx)$, and extend u_n' to the rest of the interval $[0,1]$ as a periodic function. Then the energy for all of these deformations is the same, $E[f_n] = E[f_1]$. But on $[0,1/n]$, $u_n(x) = (1/n)u_1(nx)$, and u_n is periodic, so $u_n(x)$ approaches zero pointwise as n increases. Thus the sequence of deformations is a crinkle approaching the carrier $\overline{f} = Fx$, and the limiting energy is just the energy $E[f_1]$ of the basic function used in constructing the crinkle.

In three dimensions, small replicas of a displacement field $\underline{u}$ in a body B_1 can be embedded in another body B_2 (or the same body), and a sequence of deformations approaching the uniform state can be constructed in this way, with limiting energy equal to that of $\underline{u}$ in B_1. Then the average energies that can be attained are independent of the body shape, and thus so is W^*, the infimum of these average energies.

If we regard any given non-uniform deformation as the carrier of a crinkle, the energy density at any place can relax locally through a crinkle to the value $W^*[\underline{F}(\underline{x})]$. Dacorogna [4] has shown that for problems with inhomogeneous boundary conditions (i.e. $\underline{y}_0(\underline{x}) \neq \underline{F}x$), the infimum of E is the same as the infimum of E^*, where E^* is the functional obtained by replacing W by W^*. Moreover, for

E^* a minimizer necessarily exists. The minimizer is the carrier of a crinkle that is a minimizing sequence for E.

If $W^* = W$ for all $\underline{F}$ then we say that the material is stable. Morrey [3] called this property of W _quasiconvexity_, and showed that E is lower semicontinuous for all B if and only if W is quasiconvex. Dacorogna [4] showed that whether or not W is quasiconvex, the function W^* derived from it _is_ quasiconvex.

This all suggests to me that strain energy densities deduced from experimental data will necessarily be quasiconvex, because it is the function W^* that will exhibit itself in stable equilibrium states.

The kinds of crinkles used by Morrey and Dacorogna in their proofs are more complicated than those we have discussed earlier, which in three dimensions would involve deformation gradients that are discontinuous across a one-parameter family of surfaces. Relaxation over the latter kind of crinkle leads from W to a function W_c that is said to be rank-one convex [9]. Since this is a more restricted kind of relaxation, $W > W_c > W^*$. Quasiconvexification of W_c gives W^* and rank-one convexification of W^* gives W^* back again, since rank-one crinkles are among those allowed in getting down to W^*.

For some more specific form of W it might happen that $W_c = W^*$, but it is not known that relaxation through a one-dimensional crinkle is always enough. It would be convenient if true, because rank-one convexity can be stated as an algebraic condition. Algebraic conditions necessary and sufficient for quasiconvexity are not known. Ball [9] has investigated this sort of thing, and has found various special forms of W for which quasiconvexity can be guaranteed.

<u>Acknowledgement</u>

This paper was prepared under a grant DMS-8403196 from the National Science Foundation. I gratefully acknowledge this support.

References

[1]   Young, L.C., _Lectures on the Calculus of Variations and Optimal Control Theory_. W.B. Saunders Co., Philadelphia, 1969.

[2]   Pipkin, A.C., Continuously distributed wrinkles in fabrics. Forthcoming.

[3]   Morrey, C.B., Quasi-convexity and the lower semicontinuity of multiple integrals. Pacific J. Math. $\underline{2}$, 25-53 (1952).

[4]   Dacorogna, B., Quasiconvexity and relaxation of nonconvex problems in the calculus of variations. J. Functional Anal. $\underline{46}$, 102-18 (1982).

[5]   Rivlin, R.S., Plane strain of a net formed by inextensible cords. Arch. Rat. Mech. Anal. $\underline{4}$, 951-74 (1955).

[6]   Pipkin, A.C., Some developments in the theory of inextensible networks. Quart. Appl. Mah. $\underline{38}$, 343-55 (1980).

[7]   Pipkin, A.C., Inextensible networks with slack. Quart. Appl. Math. $\underline{40}$, 63-71 (1982).

[8]   Pipkin, A.C., Energy minimization for nets with slack. Quart. Appl. Math. Forthcoming.

[9]   Ball, J.M., Convexity conditions and existence theorems in nonlinear elasticity. Arch. Rat. Mech. Anal. $\underline{63}$, 337-403 (1977).

MICROSTRUCTURES AND PHYSICAL PROPERTIES OF COMPOSITES

Ping Sheng

Corporate Research Science Laboratories
Exxon Research and Engineering Company
Clinton Township, Annandale, New Jersey   08801

Abstract

A question that has often been raised is: "How accurate is the effective medium approximation?" The question is significant in view of the fact that different effective medium theories, derived with the same goal of describing a "random" composite, can produce drastically different predictions. In the first part of this paper I illustrate with several examples that different versions of effective medium theories are actually associated with different underlying microstructures. This fact explains a major part of the discrepancies in the predictions of various effective medium theories. The recognition of the role of microstructure naturally raises to the forefront the need for a general and precise method for incorporating structural information in the calculation of electric and elastic properties of composites. The second half of the paper addresses part of this problem by presenting a first-principle approach to the calculation of effective elastic moduli for arbitrary periodic composites. By using Fourier coefficients of the periodic system as structural inputs, the new method offers the advantage of circumventing the need for explicit boundary-conditions matching across material interfaces. As a result, it can handle complex unit cell geometries just as easily as simple cell geometries.

I. Introduction

In recent years the interest in the physical properties of heterogeneous composites has focused renewed attention on the theoretical calculation of effective dielectric constants and elastic moduli for a composite medium[1,2]. At present, the effective medium (EM) theories[1,2] constitute the most prevalent approach to the problem. (The terminology "effective medium theory" has been used in widely different contexts, but in this paper it will be used to denote only those theories derivable from the coherent potential approximation.) The EM theories are generally simple to use. However, the fact that there is more than one version of the theory could sometimes be confusing. This is especially the case when different effective medium theories, derived with the same aim of describing a "random" composite, predict diverging physical characteristics. It is the purpose of the first part of this paper, Section II, to point out that while microstructure, i.e., the shapes and topological arrangements of the constituent phases, is usually not explicitly considered in the original derivations of the EM theories, each theory is nevertheless associated with an implied underlying structure for the random composite. By illustrating this association for four proto-type random composite microstsructures and their respective effective medium theories, it will be seen that many of the differing predictions for the various EM theories can be understood in terms of their different underlying geometries. The recognition of the role of microstructure naturally raises the question of whether there are more precise and general ways to incorporate such information in the effective dielectric constant and elastic moduli calculations. The second half of the paper, Section III, addresses part of this problem by presenting a new first-principle approach[3,4] to the calculation of effective elastic moduli for periodic composites with arbitrary unit cell microgeometry. The method uses the Fourier coefficients of a periodic system as structural inputs and offers the advantage of incorporating the boundary conditions implicitly in the equations of motion, thereby circumventing the traditional difficulty of matching boundary conditions across complex material interfaces. The possibility of using the new approach to calculate the properties of a random composite is discussed in Section IV.

II. Effective Medium Theories

There are a variety of ways by which one can derive the effective medium theories. Here I will adopt an approach in which the role of microstructure can be most easily delineated. Basically, the approach is based on considering the composite as made up of elementary structural units[5]. For example, in Figure 1a, we show that in a composite consisting of dispersed inclusions in a matrix the basic unit may be taken as a coated grain. If the inclusions are allowed to touch, however, then the two phases should be considered on an equal basis. This implies that a grain of constituent 1 and a gran of constituent 2 are the two basic units. The resulting structure is schematically illustrated in Figure 1b. A cluster of grains may also be considered as a basic structural unit, but then the theory would lose its calculational simplicity. Once the basic units are chosen, the next step is the embedding of each individual unit in a homogeneous effective medium characterized by a yet undetermined effective complex dielectric constant $\bar{\varepsilon}$ and elastic moduli $\bar{\mu}, \bar{\lambda}$. To calculate $\bar{\varepsilon}$, the forward scattering amplitude $f_i(0)$ for an incident eletromagnetic plane wave by the $\underline{ith}$ unit is calculated in the long wavelength limit. The effective medium condition is then[5-7]

$$\sum_i v_i f_i(0) = 0, \qquad (1)$$

where v_i is the volume fraction of the $\underline{ith}$ unit. Since $f_i(0)$ is a function of $\bar{\varepsilon}$, Equation (1) represents a condition for its determination. For the elastic constants, on the other hand, the equations are:

$$\sum_i v_i f_i^{P}(0) = 0, \qquad (2a)$$

and

$$\sum_i v_i f_i^{S}(0) = 0, \qquad (2b)$$

where $f_i^{p(s)}(0)$ (a function of $\bar{\lambda}, \bar{\mu}$) is the forward scattering amplitude of the p(s) elastic plane wave by the $\underline{ith}$ embedded unit evaluated at the long wavelength limit.

The condition that the forward scattering amplitudes must be zero on the average can be justified heuristically as follows. Since the wavelength of the probing wave is much longer than the scale of basic structural units, the inhomogeneities cannot be individually resolved. That is, the medium should appear homogeneous to the probing wave. Therefore, if one looks in the propagating direction of a plane wave there should be, on the average, no net scattering out of the beam. Application of this approach for deriving the EM theories shows that the input structural units determine the type of theory obtained. For example, we can have two types of random composites with the microstructures depicted schematically in Figure (1a) and (1b). By using for the basic structural unit a coated sphere with a coating layer thickness determined by the overall composition, we get the equation for the effective dielectric constant $\bar{\varepsilon}$ of the dispersed inclusion microstructure:

$$\frac{\bar{\varepsilon} - \varepsilon_m}{\bar{\varepsilon} + 2\varepsilon_m} + p \frac{\varepsilon_m - \varepsilon_i}{\varepsilon_i + 2\varepsilon_m} = 0, \tag{3}$$

where ε_m is the matrix dielectric constant, ε_i the inclusion dielectric constant, and p the inclusion volume fraction. Equation (3) is known as the Maxwell-Garnett theory[8]. For the symmetric microgeometry (Figure 1b), on the other hand, there are two structural units (spheres of two components) as mentioned before. The resulting equation is

$$p \frac{\bar{\varepsilon} - \varepsilon_1}{2\bar{\varepsilon} + \varepsilon_1} + (1 - p) \frac{\bar{\varepsilon} - \varepsilon_2}{2\bar{\varepsilon} + \varepsilon_2} = 0, \tag{4}$$

which is known in the literature as Bruggeman's effective medium theory[9]. The predictions of Equations (3) and (4) are very different. For dc conductivity of a metal-insulator composite, Equation (4) predicts a percolation threshold at one-third volume fraction of metal (i.e., conductivity = 0 for metal fraction less than one-third), whereas Equation (3) tells us that for insulating inclusions, the dc conductivity vanishes only when metal volume fraction approaches zero. This difference is easily linked with the differing microstructures. For the dispersed inclusion microstructure, the insulator can never fully block off the conducting matrix unless it is at $p = 1$. However, for the symmetric microgeometry the two constituents can undergo a matrix inversion as their relative volume fraction is varied. This accounts for the percolation threshold. At the optical frequency regime (but wavelength still $\gg$ scale of inhomogeneities) the effective dielectric constants calculated from Equations (3) and (4) display equally different behaviors. In Figure 2 we show the predictions of the two theories compared with experimental results[10] (on the real and imaginary parts of index of refraction, $n + ik$) for two composites with microgeometries which can be approximated by that of Figures (1a) and (1b). It is seen that while the agreement is not perfect, there is obvious general accord between theory and experiment. The differences in the two cases, which cannot be accounted for by the constituents' material properties, offer a clear demonstration of the microstructural effect.

For the elastic properties of these two microstructures, we have solved (with the help of MIT's symbolic manipulation program MACSYMA) the general problem of elastic wave scattering from a coated sphere embedded in an arbitrary medium and obtained explicit expressions for the forward scattering amplitudes in the long wavelength limit. In the case of the dispersed inclusion microgeometry, we get[11]

$$\bar{\beta} = \frac{\beta_m [4(\mu_m - \mu_i) + 3\beta_i]}{[4(\mu_m - \mu_i) + 3\beta_i](1 - p) + 3p\beta_m} - \frac{4}{3}(\mu_m - \bar{\mu}) \tag{5a}$$

$$A\left(\frac{\bar{\mu}}{\mu_m}\right)^2 + B\left(\frac{\bar{\mu}}{\mu_m}\right) + C = 0, \tag{5b}$$

where $\bar{\beta} = \bar{\lambda} + 2\bar{\mu}$,

$$A = -160(3k_2 + 2)R_{10}p^{10/3} - 2000\,R_7\,p^{7/3} + 10080\,R_5\,p^{5/3} - 2000$$
$$(3k_2^2 - 4k_2 + 3)R_3 p - 80(2k_2 + 3)R_0, \tag{5c}$$

$$B = -80(3k_2 - 8)R_{10}p^{10/3} + 4000\,R_7 p^{7/3} - 20160\,R_5 p^{5/3} + 1500$$
$$(5k_2^2 - 14k_2 + 8)R_3 p - 30(3k_2 - 16)R_0 \tag{5d}$$

$$C = 80(9k_2 - 4)R_{10}p^{10/3} - 2000\,R_7 p^{7/3} + 10080\,R_5 p^{5/3} - 250$$
$$(27k_2^2 - 52k_2 + 24)R_3 p + 10(19k_2 - 24)R_0, \tag{5e}$$

and

$$R_{10} = (k_5 - 1)(9k_2 k_5 - 4k_5 + 6k_2 + 4)(38k_2 k_3 k_5 - 48k_3 k_5 + 57k_2 k_5 - 75k_5$$
$$- 38k_2 k_3 - 57k_3 + 48k_2 + 72),$$

$$R_7 = (9k_2 k_5 - 4k_5 + 6k_2 + 4)(54k_2^2 k_3 k_5^2 - 104k_2 k_3 k_5^2 + 48k_3 k_5^2 + 81k_2^2 k_5^2$$

$$- 156k_2 k_5^2 + 72k_5^2 + 3k_2^2 k_3 k_5 - 35k_2 k_3 k_5 + 9k_3 k_5 - 90k_2^2 k_5 + 252k_2 k_5 -$$

$$- 144k_5 - 57k_2^2 k_3 + 76k_2 k_3 - 57k_3 + 72k_2^2 - 96k_2 + 72),$$

$$R_5 = (k_2 - 1)^2 (k_5 - 1)(9k_2 k_5 - 4k_5 + 6k_2 + 4)(16k_3 k_5 + 24k_5 + 19k_3 - 24),$$

$$R_3 = R_5/(k_2 - 1)^2,$$

$$R_0 = (9k_2 k_5 - 4k_5 + 6k_2 + 4)^2 (16k_3 k_5 + 24k_5 + 19k_3 - 24),$$

with

$$k_2 = \frac{\beta_m}{\mu_m}, \quad k_3 = \frac{\beta_i}{\mu_i}, \quad \text{and} \quad k_5 = \frac{\mu_m}{\mu_i}.$$

In the equations above, the subscripts i and m denote the inclusion and the matrix components, respectively, and p is the volume fraction of inclusions. Equation (5a) for $\bar{\beta}$ is exactly equivalent to one of the Hashin-Shtrikman bounds[12], and Equation (5b) turns out to be identical to that derived by Christensen and Lo[13]. For the elastic constants of a composite with symmetric microstructure, the scattering amplitudes of spheres can be obtained from the coated sphere case by letting the coating thickness approach zero. Equations (2a) and (2b) then become

$$\frac{1}{\bar{\beta}} = \frac{p}{\beta_1 + \frac{4}{3}(\bar{\mu} - \mu_1)} + \frac{1 - p}{\beta_2 + \frac{4}{3}(\bar{\mu} - \mu_2)}, \tag{6a}$$

$$\frac{1}{\bar{\mu} + H} = \frac{p}{\mu_1 + H} + \frac{1 - p}{\mu_2 + H}, \tag{6b}$$

and

$$H = \bar{\mu}\, \frac{9\bar{\beta} - 4\bar{\mu}}{6\bar{\beta} + 4\bar{\mu}}. \tag{6c}$$

Here p denotes the volume fraction of component 1. Equations (6a)-(6c) were first derived by Berryman[14]. In Figures 3 and 4 the predictions of Equations (5) and (6) are compared for a fluid-solid composite. The quantities calculated are $\bar{\mu}$, $\bar{\beta}$, and the $p(s)$ wave attenuation, $1/Q_{\bar{\beta}(\bar{\mu})} = \mathrm{Im}[\bar{\beta}(\text{or } \bar{\mu})]/\mathrm{Re}[\bar{\beta} \text{ (or } \bar{\mu})]$ as a function of water concentration p. The values of the parameters used in the numerical computation are: for solid, $\beta_s = 8 \times 10^{11}$ dynes/cm^2 and $\mu_s = 2 \times 10^{11}$ dynes/cm^2; for water, $\beta_f \quad \kappa + i\,\omega\,(\lambda_\ell + 2\eta)$, $\mu_f = i\omega\eta$, where $\kappa = 2.2 \times 10^{10}$ dynes/cm^2 is the bulk modulus of water, $\eta = 1$ centipoise is the viscosity of water, $\lambda_\ell = -2\eta/3$ is the bulk viscosity, and ω is the angular frequency, taken to be 10^6 rad/sec in the present calculation. It is seen that the moduli for the dispersed inclusion (water being the inclusion) microgeometry decrease slower than that of the symmetric microgeometry case, which has a rigidity threshold at water fraction $p = 0.6$. Physically, the threshold corresponds to the breakup of the solid frame at the point of matrix inversion, which cannot occur in the dispersed inclusion microgeometry. The difference in attenuation between the two microstructures is also seen to be striking. The p-wave attenuation $(1/Q_\beta)$ in the symmetric microgeometry case is a factor 2 to 10^3 times higher than that of the dispersed inclusion microgeometry. This can be understood physically by observing that whereas water in the dispersed inclusion microgeometry is confined within spherical pores and immobile, the symmetric microgeometry allows much larger fluid movements in its connected pores, thereby increasing the viscous attenuation.

Another form of random composite microstructure[5], found in a large number of sputtered or evaporated cermet films, can be modeled by two coated particle structural units (of the same relative composition) in which the material constituents for the coating and the grain in one unit have the reversed roles in the other. If we consider the grains to be spheroidal in shape, then the resulting effective medium equations are[5]

$$fD_1 + (1 - f)D_2 = 0 \tag{7a}$$

$$D_1 = \frac{2}{3}\, D[\bar{\epsilon}, \epsilon_1, \epsilon_2, p, A(\alpha,u), B(\alpha)] + \frac{1}{3}\, D[\bar{\epsilon}, \epsilon_1, \epsilon_2, p, 3-2A(\alpha,u), 3-2B(\alpha)], \tag{7b}$$

$$D_2 = \frac{2}{3}\, D[\bar{\epsilon}, \epsilon_2, \epsilon_1, 1-p, A(\beta,v), B(\beta)] + \frac{1}{3}\, D[\bar{\epsilon}, \epsilon_2, \epsilon_1, 1-p, 3-2A(\beta,v), 3-2B(\beta)], \tag{7c}$$

$$f = \frac{(1 - p^{1/3})^3}{(1 - p^{1/3})^3 + [1 - (1 - p)^{1/3}]^3}. \tag{7d}$$

Here the functional form of f, the relative abundance of structural unit 1, is determined by the kinetics of the film formation process, p is the volume fraction of constituent 1, α is the ratio between the minor (major) and major (minor) axes of the elliptic cross section for oblate (prolate) spheroidal grain in unit 1, β is the similar quantity for the grain in unit 2, $u = (p/\alpha)^{1/3}$, and

$v = [(1 - p)/\beta]^{1/3}$. The functional forms of D, A, and B are:

$$D[\bar{\epsilon},x,y,\mu,A,B] = \frac{[A\bar{\epsilon} + (3 - A)y](y - x)\mu + [Bx + (3 - B)y](\bar{\epsilon} - y)}{A(3 - A)(\bar{\epsilon} - y)(y - x)\mu + [Bx + (3 - B)y][Ay + (3 - A)\bar{\epsilon}]} , \tag{8a}$$

$$A(\gamma,\omega) = \frac{3}{2}\frac{1}{(1 - \gamma^2)\omega^3}\left[\frac{1}{(1 - \gamma^2)^{1/2}}\tan^{-1}\frac{(1 - \gamma^2)^{1/2}\omega}{(s^2 + \gamma^2\omega^2)^{1/2}} - \omega(s^2 + \gamma^2\omega^2)\right], \tag{8b}$$

$B(\gamma) = A(\gamma,1/\gamma^{1/3})$ evaluated at $s = 0$, s being the solution of the equation $(s^2 + \omega)^2(s^2 + \gamma^2\omega^2) = 1$. The predictions of Eqs. (7) and (8) have some similarities to both the dispersed-inclusion and symmetric microstructure cases. As shown in Figure 5, the dc conductivity exhibits a percolation threshold as in the symmetric case. However, the (grain shape dependent) curvature in the conductivity vs. metal fraction p curve and the value of the threshold make the present theory distinct from Bruggeman's theory and at the same time offers a good account for the behavior of the experimental data. At optical frequencies the present theory has some qualitative features, such as the absorption peaks seen in Figure 6, which are in common with the dispersed inclusion case. However, both the position and width of the peak and the infrared behaviors are different from the predictions of the dispersed-inclusion microstructure. Elastic properties of the cermet microstructure can also be calculated from the forward scattering amplitudes of coated elastic grains. Numerical results are presently being generated.

A fourth example of the random composite microstructure is that belonging to a general class of approaches called differential effective medium (DEM) theories[15-22]. The basic process of building up a "differential effective medium" material is illustrated in Figure 7. Starting with a homogeneous component 1, a small fraction is replaced by component 2. The resulting effective medium property can be calculated as:

$$F(\bar{\epsilon},1 - \Delta\psi, \epsilon_1, \Delta\psi,\epsilon_2) = 0, \tag{9}$$

where $F = 0$ represents the effective medium condition. Differentiation of Equation (9) yields the differential parameter changes induced by the replacement step:

$$\left(\frac{\partial F}{\partial \bar{\epsilon}}\right)_{\psi=0} \frac{d\bar{\epsilon}}{d\psi} = -\left(\frac{\partial F}{\partial \psi}\right)_{\psi=0} , \tag{10}$$

or

$$d\bar{\epsilon} = -\frac{(\partial F/\partial \psi)_{\psi=0}}{(\partial F/\partial \bar{\epsilon})_{\psi=0}} d\psi. \tag{11}$$

It should be remarked that Equation (11) is independent of the choice of F because all the EM theories agree to first order in the dilute limit. By regarding the resulting composite as a homogeneous medium, we can iterate the replacement process and thereby build up the relative concentration of component 2. However, since the composite already has some component 2, the replacement would increase the fraction of component 2 by only

$$d\psi_{actual} = (1 - \psi_{actual})d\psi. \tag{12}$$

Therefore, the final equation for the DEM is

$$d\bar{\epsilon} = -\frac{(\partial F/\partial \psi)_{\psi=0}}{(\partial F/\partial \bar{\epsilon})_{\psi=0}} \frac{d\psi}{(1 - \psi)}. \tag{13}$$

In the case of spherical replacement units, Equation (13) can be readily integrated with the initial condition $\bar{\epsilon} = \epsilon_1$ at $\psi = 0$:

$$1 - \psi = \left(\frac{\bar{\epsilon}/\epsilon_1 - \epsilon_2/\epsilon_1}{1 - \epsilon_2/\epsilon_1}\right)\left(\frac{\epsilon_1}{\bar{\epsilon}}\right)^{1/3} . \tag{14}$$

The effective elastic moduli equations can be obtained similarly. In case of spherical units, the coupled differential equations are[18]

$$d\bar{\beta} = F^{\beta}(\bar{\beta},\bar{\mu},\beta_2,\mu_2) \frac{d\psi}{1 - \psi} , \tag{15a}$$

$$d\bar{\mu} = F^{\mu}(\bar{\beta},\bar{\mu},\beta_2,\mu_2) \frac{d\psi}{1 - \psi} , \tag{15b}$$

where $\quad F^{\beta} = 3\bar{\beta}\,\dfrac{8(\mu_2-\bar{\mu})[4\bar{\mu}(\mu_2-\bar{\mu}) + \bar{\beta}(\mu_2+2\bar{\mu}) - 3\bar{\mu}\beta_2] + 3\beta(\bar{\beta}-\beta_2)(2\mu_2+3\bar{\mu})}{(4\mu_2 - 4\bar{\mu} - 3\beta_2)(4\bar{\mu}\mu_2 + 6\bar{\beta}\mu_2 - 4\bar{\mu}^2 + 9\overline{\mu\beta})} , \tag{15c}$

$$F^{\mu} = \frac{15\overline{\beta\mu}(\mu_2 - \bar{\mu})}{4\bar{\mu}\mu_2 + 6\bar{\beta}\mu_2 - 4\bar{\mu}^2 + 9\overline{\mu\beta}} , \tag{15d}$$

(Notice that F^μ is actually independent of β_2 in this case.) The initial conditions are $\bar{\beta} = \beta_1$, $\bar{\mu} = \mu_1$, at $\psi = 0$. A possible realization of the DEM microstructure is shown in Figure 8. The essential characteristic is the diversity of grain sizes. This arises from the fact that at each replacement step the composite has to look homogeneous to the replacing grains, which is possible only if the grains are increasing in size at each succeeding replacement step. Another feature of the DEM microstructure, as pointed out by Yonezawa and Cohen[23], is that the starting component will always remain connected and the replacing component always disconnected (unless $\psi = 1$). In this aspect it is similar to the dispersed inclusion microstructure.

Physical properties of the DEM microgeometry, as calculated from the relevant mathematical equations, offer a surprisingly good description for some of the electrical and acoustical characteristics of sedimentary rocks[17,18]. For the electrical properties, the fluid-saturated sedimentary rocks are known to exhibit definite correlation between the dc conducitivity σ and the porosity ϕ. The empirical relation[24] $\sigma = \sigma_f \phi^m$, where σ_f is the fluid conductivity and $m \simeq 2$ is a constant, is known as Archie's law. Viewed in the framework of percolation theory, Archie's law implies that sandstone is a random composite with a structure that yields a percolation threshold $\phi = 0$. This aspect agrees with both the DEM and the dispersed inclusion microgeometries. However, the fact that the exponent $m > 1$ makes Archie's law distinct from the predictions of any of the previous microstructures, which always give $m = 1$. The physical meaning of $m > 1$ is that not all the pores contribute equally to the conductivity and that only a subset of porosity is effective in determining the overall conductivity of the composite. If we examine the prediction of the DEM microstructure, it turns out that it yields exactly the same form of Archie's law, $\sigma = \sigma_f \phi^m$, where m is a function of the grain shape. For spheres $m = 3/2$, and $m = 2$ for oblate spheroids with aspect ratio $\simeq 4.5$. The success of the DEM theory in explaining Archie's law naturally raises the curiosity as to how well it can predict rock's acoustical properties. It turns out that recently, great interest has been aroused by sonic attenuation experiments on rocks in which the fluid-saturated sandstones are shown to exhibit a frequency peak in attenuation whose value is up to several orders of magnitude larger than the attentuation of either the fluid or the solid phase alone[25]. In Figure 9 we compare[17,18] experimental results with the theoretical predictions of DEM equations using the appropriate material parameters and spheroidal replacement units (with aspect ratio 4.5 so as to be consistent with Archie's law). It is seen that both the magnitude and width of the attenuation peak, together with the associated velocity dispersion, can be accounted for by the DEM. The small value of porosity, 0.3%, used in the calculation to fit the data gives an indication that the attenuation is probably associated with cracks. This is backed by results of pressure experiments[26] in which the closing of cracks by the applications of pressure is shown to result in drastically reduced attenuation with only a small accompanying decrease in porosity. Also, the fact that the experimental data, which were taken by using either water (open circles) or glycerine (filled circles) as the filling fluid, seem to fall on a single curve as a function of ηf (product of viscosity and frequency) lends strong support to the theoretical picture of attenuation by viscous dissipation in the fluid. From this general evidence it follows that the large magnitude of the attenuation is understandable in terms of the amplification of strain[17,18] in the microcracks since the fluid-filled cracks are more compressible than the solid. The increased elastic energy density in a dissipative medium naturally gives rise to enhanced attenuation.

The discussion of four microstructures and their respective physical properties as predicted by the EM theories has shown that the often asked question, "How accurate is the effective medium approximation?", should really be rephrased as, "How accurate is the effective medium approximation relative to its implied microstructure?" But even that may not be satisfactory, since it has been shown[27] that there exist composite geometries for which the predictions of effective medium theories are rigorously valid. Therefore, for a given EM theory the true question is, "What is the microstructure?" However, the recognition of the central role of microstructure immediately leads to the realization that since there are only a finite number of EM theories, there is no possibility that they

can exhaust all possible microstructures. In fact, all EM theories have rather
simple structural units as building blocks. In order to consider complicated unit
cell geometries, it is necessary to develope more general, practical method which
is capable of calculating the properties of composite with arbitrary unit cell
microstructure. This is the subject addressed in the next section.

III. First-Principle Approach

In the traditional approach to effective constants calculation the most difficult step usually lies in the matching of boundary conditions across material
interfaces. Since the procedure is highly sensitive to the shape of the interfaces,
only relatively simple unit cell geometries have been considered. However, this
difficulty can be circumvented by directly considering the equations of motion for
the inhomogeneous medium. For elastic systems, this means the inhomogeneous
elastic wave equation:

$$\rho(\vec{r})\ddot{\vec{u}}(\vec{r}) = [\lambda(\vec{r}) + \mu(\vec{r}))] \, \nabla \, (\nabla \cdot \vec{u}) + \mu(\vec{r})\nabla^2 \, \vec{u} + [(\nabla\cdot\vec{u})\nabla\lambda(\vec{r}) +$$

$$\nabla\mu(\vec{r}) \cdot (\nabla\vec{u}) + (\nabla\vec{u}) \cdot \nabla\mu(\vec{r})], \tag{16}$$

where $\rho(\vec{r})$ is the density, $\vec{u}(\vec{r})$ is the displacement field, and $\lambda(\vec{r})$, $\mu(\vec{r})$ are
the spatially varying elastic constants. For an abrupt material interface, $\lambda(\vec{r})$
and $\mu(\vec{r})$ approach a step function. The classical elasticity boundary conditions, displacements and tractions continuous across the interface, can be shown
to result directly from Equation (16)[4]. It follows that the solution of Equation
(16) should yield complete dynamical information about the system. For the calculation of effective moduli (which are defined in the long wave-length limit) of a
periodic system, it is much simpler to deal with the Fourier transform of Equation
(16):

$$\sum_{\beta} [E^2 \delta_{\alpha\beta} - \omega^2_{\alpha\beta}(\vec{k})]u_{\beta}(\vec{k}) = \sum_{\vec{K}_n \neq 0} [\sum_{\beta} \frac{1}{\rho(0)} V_{\alpha\beta}(\vec{k},\vec{K}_n) \, u_{\beta}(\vec{k} - \vec{K}_n)$$

$$- E^2 \frac{\rho(\vec{K}_n)}{\rho(0)} \, u_{\alpha}(\vec{k} - \vec{K}_n)]. \tag{17a}$$

Here we have assumed the time dependence of $\vec{u}$ is $\exp[-iEt]$, $\alpha,\beta = 1,2,3$ denote the three components of a vector, $\vec{k}$ is a continuous wavevector, $\vec{K}_n$ is the nth reciprocal lattice vector of the periodic structure (for three dimensional structures $n \equiv \{n_1,n_2,n_3\}$, where the three numbers index the periodicities along the three principal axes), and

$$V_{\alpha\beta}(\vec{k},\vec{K}_n) = \begin{cases} (\lambda(\vec{K}_n) + \mu(\vec{K}_n))k_\alpha(k_\alpha - K_{n\alpha}) + \mu(\vec{K}_n)\vec{k}\cdot(\vec{k} - \vec{K}_n), & \alpha = \beta \\ \lambda(\vec{K}_n)k_\alpha(k_\beta - K_{n\beta}) + \mu(\vec{K}_n)\,k_\beta(k_\alpha - K_{n\alpha}), & \alpha \neq \beta, \end{cases} \tag{17b}$$

$$\omega^2_{\alpha\beta}(k) = \frac{1}{\rho(0)}\,V_{\alpha\beta}(\vec{k},0). \tag{17c}$$

It should be remarked that $V_{\alpha\beta}(\vec{k},\vec{K}_n)$, $\vec{K}_n \neq 0$, simply represents the effect of multiple scatterings by the elastic constant inhomogeneities. For a homogeneous medium, right hand side of Equation (17a) vanishes, and it is straighforward to check that the diagonalization of the left hand side directly yields the eigenfrequencies of acoustic and shear waves.

If we now specialize to the case of elastic waves in the limit of $|\vec{k}| \to 0$, with $E \propto |\vec{k}|$, then an analysis of the order of magnitude of the various terms show that the $E^2 \rho(\vec{k}_n)u_\alpha(\vec{k} - \vec{K}_n)/\rho(0)$ $(\vec{k}_n \neq 0)$ term is higher order in $|\vec{k}|$ $(|\vec{k}|^3)$ than the other terms (which are $|\vec{k}|^2$) and therefore can be neglected in the calculation of effective moduli. Physically, the absence of density effects (apart from $\rho(0)$) is expected since the effective moduli measures only the static potential energy of the system, which is independent of ρ. If now for $\vec{K}_n \neq 0$ we define a scattering matrix $S(\vec{k},\vec{K}_n)$ such that

$$\vec{u}(\vec{k} - \vec{K}_n) = S(\vec{k},\vec{K}_n)\vec{u}(\vec{k}), \tag{18}$$

then Equation (17a) becomes

$$\sum_\beta [v^2\delta_{\alpha\beta} - C^2(\vec{k}) - \tau_{\alpha\beta}(\vec{k})]u_\beta(\vec{k}) = 0, \tag{19}$$

where $v^2 \equiv E^2/|\vec{k}|^2$, $C^2_{\alpha\beta(\vec{k})} = \omega^2_{\alpha\beta}(\vec{k})/|\vec{k}|^2$, and

$$\tau_{\alpha\beta}(\vec{k}) = \frac{1}{\rho(0)} \sum_{K_n \neq 0} \sum_n |\vec{k}|^{-2} v_{\alpha n}(\vec{k},\vec{K}_n)\, S_{n\beta}(\vec{k},\vec{K}_n). \qquad (20)$$

Diagonalization of Equation (19) yields directly the eigen velocities v of the inhomogeneous structure. The analysis of the associated eigenfunctions could then tell us about the polarizations of the elastic wave eigenmodes. For a longitudinal wave in a cubic structure $(\vec{k} \parallel [1,0,0])$, for example, we define $\overline{K}_{11} = v^2 \rho(0)$. Other moduli can be defined similarly.

From the definition of $\hat{S}(\vec{k},\vec{K}_n)$, Equation (18), it can be shown after straightforward but lengthy algebra that $\hat{S}$ rigorously satisfies the matrix Dyson's equation for elastic wave scattering:

$$\hat{S}(\vec{k},\vec{K}_n) = \hat{F}(\vec{k} - \vec{K}_n,\vec{K}_n) + \sum_{\substack{\vec{K}_n \neq \vec{K}_n \\ n \neq 0}} \hat{F}(\vec{k} - \vec{K}_n,\vec{K}_n - \vec{K}_n')\hat{S}(\vec{k},\vec{K}_n'). \qquad (21a)$$

Here the elements of $\hat{F}$ are defined by

$$F_{\alpha\beta}(\vec{k} - \vec{K}_n,\vec{K}_n - \vec{K}_n') = \frac{1}{1 + \gamma_\alpha(\vec{k} - \vec{K}_n)} \frac{1}{E^2 - \omega_{\alpha\alpha}^2(\vec{k} - \vec{K}_n)} \{V_{\alpha\beta}(-\vec{k} + \vec{K}_n,\vec{K}_n - \vec{K}_n')$$

$$+ A \frac{(k - K_n)_\alpha}{1 - \delta(\vec{k} - \vec{K}_n)} \sum_n \frac{(\vec{k} - \vec{K}_n)_n}{1 + \gamma_n(\vec{k} - \vec{K}_n)} \frac{V_{n\beta}(-\vec{k} + \vec{K}_n,\vec{K}_n - \vec{K}_n')}{E^2 - \omega_{nn}^2(\vec{k} - \vec{K}_n)} \}, \qquad (21b)$$

where $A = (\lambda(0) + \mu(0))/\rho(0)$,

$$\gamma_\alpha(\vec{k} - \vec{K}_n) = A \frac{(\vec{k} - \vec{K}_n)_\alpha^2}{E^2 - \omega_{\alpha\alpha}^2(\vec{k} - \vec{K}_n)}, \qquad (21c)$$

and

$$\delta(\vec{k} - \vec{K}_n) = \sum_\alpha \frac{\gamma_\alpha(\vec{k} - \vec{K}_n)}{1 + \gamma_\alpha(\vec{k} - \vec{K}_n)}. \qquad (21d)$$

Equation (21), with E^2 neglected in comparison to $\omega_{\alpha\alpha}^2(\vec{k} - \vec{K}_n)$ $(\vec{K}_n \neq 0)$, can be solved iteratively. Rapid numerical convergence was observed. In the worst case (spheres just touching, for example) convergence to three or four significant figures was obtained after 15 to 20 iterations. Physically, the iterations can be interpreted as multiple scatterings of the plane wave by structural inhomogeneities.

The combination of using Fourier coefficients of material parameters as inputs and the capacity to harness the iterative solution technique make the approach described above generally applicable to the elastic moduli calculation of periodic composites with arbitrary unit cell geometry. Here the results on two examples are presented. In the first case the elastic moduli of a two-dimensional fractal structure, the Sierpinski carpet, are calculated to three stages[3]. The geometries of the unit cell for the three stages are shown in Figure 10. By solving Equation (21) for up to a maximum $\pm N$ of K_n's for each of the two directions, with $\lambda + 2\mu = 1$ and $\mu = 0.4$, we obtain the effective moduli as functions of N. In Figure 11a, values of $\overline{K}_{11}/\rho(0)$ for stages 1, 2, and 3 are plotted as a function of $1/N$. The fact that the points lie on good straight lines is utilized to extrapolate the results to $1/N = 0$. In Figure 11b we show $\log(\overline{K}/\rho(0))$ as a function of $\log L$, where the size of the fractal L is defined to be the ratio between the sides of the largest and the smallest squares at each stage. The excellent power-law behaviors demonstrate that $\overline{K}_i/\rho(0) \propto L^{-\alpha_i}$, where $\overline{K}_1 \equiv \overline{K}_{11}$ (longitudinal wave in [10] direction), $\overline{K}_2 \equiv \overline{K}_{11} - \overline{K}_{12}$ (shear wave in [11] direction), and $\overline{K}_3 \equiv \overline{K}_{44}$ (shear wave in [10] direction) are the three moduli associated with three of the eigenmodes of elastic wave excitations. The values of α are $\alpha_1 = 0.16 \pm 0.01$, $\alpha_2 = 0.14 \pm 0.01$, and $\alpha_3 = 0.35 \pm 0.01$. The error bars on the points indicate the range of possible extrapolation errors estimated by using maximum and minimum slopes compatible with the data. Calculations have also been performed using different ratios of μ/λ. The exponents obtained are the same, indicating the independence of the α_i's from material properties. This scaling behavior of the fractal structural moduli is compatible with the general power-law dependence of fractal properties on the size of the system. The fact that there are, within our calculational errors, two distinct scaling behaviors is quite interesting (it shows that the Sierpinski carpet undergoes a solid-liquid transformation as $L \rightarrow \infty$) since all the discrete elastic fractal models[28-30] display only one scaling exponent.

As another application of the present approach, the elastic moduli of a frame composed of fused solid spheres are considered. The unit cell of this simple cubic structure is what remains of a sphere after six caps of maximum height η (using length units where the side of the cell is 2) have been cleaved off so that it can fit inside the cubic cell. Cross sections of the structure are depicted in Figure 12. From the point of view of the traditional approach, this structure is difficult to calculate because the boundary conditions at the free surface and the fused interfaces are different. By using the new approach, the elastic moduli can be directly obtained by solving Equation (21) for up to $(2N + 1)$ K_n's in each direction. Figure 13 displays the calculated moduli as a function of $(2N + 1)^{-\alpha}$ for different values of η, where the value of α is determined by the best linearity of the result as a function of $(2N + 1)^{-\alpha}$. It is seen that for η large, i.e., small porosity, the results change only slightly as one varies N. However, for η small the variation of N yields significant changes in the final results. This is expected on the physical ground that as $\eta \to 0$, the contact areas between spheres decrease, and the moduli become sensitive to small-scale resolution of the structure in the neighborhood of contacts. In Figure 13 straight line extrapolation is used to obtain limiting values of the effective moduli. For $\eta = 0$, i.e., the spheres just touching, this procedure yields almost exactly $\overline{K}_{11}$, $\overline{K}_{44} = 0$, the expected results. From such checks we expect that extrapolation based on the asymptotic dependence of calculated results should yield accurate limiting values. Figure 14 displays the normalized moduli obtained in the above manner as a function of porosity.

IV. Concluding Remarks

From the examples described above, we can see that by using the new approach it is possible to calculate the effective moduli of a periodic composite with arbitrary unit cell geometry. Provided that one can use a big enough cluster as the unit cell, a good approximation to random composite properties may be obtained in this fashion. With an IBM 3033 computer (with a speed of ~1-2 megaflops), it is possible to obtain reasonably accurate results on a 5x5x5 cubic cell (with random occupation) in about ten hours. With the increasing availability of super-computers (with a speed of 100~200 megaflops), the time required for such calculations may be reduced dramatically. As another application of the present approach, effective dielectric constant of a composite can be obtained in a similar manner from Maxwell's equations with spatially varying dielectric constant. Provided that there is no magnetic permeability inhomogeneity (which is valid for a large class of composites made from non-magnetic materials), the calculational effort is about an order of magnitude less than the elastic case. Numerical work in this area is presently under way.

References

1. See, for example "Macroscopic Properties of Disordered Media", edited by R. Burridge, S. Childress, and G. Papanicolaou, (Springer-Verlag, New York 1982) and references therein.

2. A large number of articles and references on effective electrical properties of composites may be found in "Electrical Transport and Optical Properties of Inhomogeneous Media", edited by J.C. Garland, D.B. Tanner, AIP Conf. Proc. 40 (American Institute of Physics, New York, 1978).

3. P. Sheng and R. Tao, Phys. Rev. B31, 6131 (1985).

4. R. Tao and P. Sheng, J. Acoust. Soc. Am. 77, 1651 (1985).

5. P. Sheng, Phys. Rev. lett. 45, 60 (1980); P. Sheng, Optics and Laser Technology, Oct., 1981, p. 253-260.

6. M. Lax, Rev. Mod. Phys. 23, 287 (1951).

7. D. Stroud and F.P. Pan, Phys. Rev. B17, 1602 (1978).

8. J.C. Maxwell-Garnett, Phil. Trans. Roy. Soc. 203, 385 (1904).

9. D.A.G. Bruggeman, Ann. Phys. (Leipzig) 24, 636 (1935).

10. U.J. Gibson, H.G. Craighead, and R.A. Buhrman, Phys. Rev. B25, 1449 (1982).

11. A.J. Callegari, D. McGlaughlin, and P. Sheng, to be published.

12. Z. Hashin and S. Shtrikman, J. of Franklin Institute 271, 336 (1961). The same result has also been derived by E.H. Kerner, Proc. Phys. Soc. (London) B69, 808 (1956).

13. R.M. Christensen and K.H. Lo, J. Mech. Phys. Solids 27, 315 (1979). I wish to thank B. Nair for verifying the identity of the two formulae.

14. J.G. Berryman, J. Acoust. Soc. Am. $\underline{68}$, 1809 (1980).

15. See, for example, R. Landauer's review article in Ref. 2.

16. P.N. Sen, C. Scala, and M.H. Cohen, Geophysics $\underline{46}$, 781 (1981).

17. P. Sheng and A.J. Callegari, Appl. Phys. Lett. $\underline{44}$, 738 (1984).

18. P. Sheng and A.J. Callegari, AIP Conf. Proc. No. 107, "Physics and Chemistry of Porous Media" edited by D.L. Johnson and P.N. Sen, (American Institute of Physics, New York, 1984), p. 144.

19. F.S. Henyey and N. Pomphrey, Geophys. Res. Lett. $\underline{9}$, 903 (1982).

20. W.M. Bruner, J. Geophys. Res. $\underline{81}$, 2573 (1976).

21. S. Boucher, Revue M$\underline{22}$, I (1976).

22. A. Norris, P. Sheng, A.J. Callegari, J. Appl. Phys., $\underline{57}$, 1990 (1985).

23. F. Yonezawa and M.H. Cohen, J. Appl. Phys. $\underline{54}$, 2895 (1983).

24. G.E. Archie, Petroleum Technology $\underline{5}$, No. 1, (1942).

25. J.R. Bulau, B.R. Tittmann, and M. Abdel-Gawad, Trans. Am. Geophys. Union $\underline{64}$, No. 18, 324 (1983).

26. K. Winkler and A. Nur, Geophysics $\underline{47}$, 257 (1982).

27. G.W. Milton, AIP Conf. Proc. No. 107, "Physics and Chemistry of Porous Media" edited by D.L. Johnson and P.N. Sen (American Institute of Physics, New York, 1984), p. 66.

28. S. Feng and P.N. Sen, Phys. Rev. Lett. $\underline{52}$, 216 (1984).

29. Y. Kantor and I. Webman, Phys. Rev. Lett. $\underline{52}$, 1891 (1984).

30. D.J. Bergman and Y. Kantor, Phys. Rev. Lett. $\underline{53}$, 511 (1984).

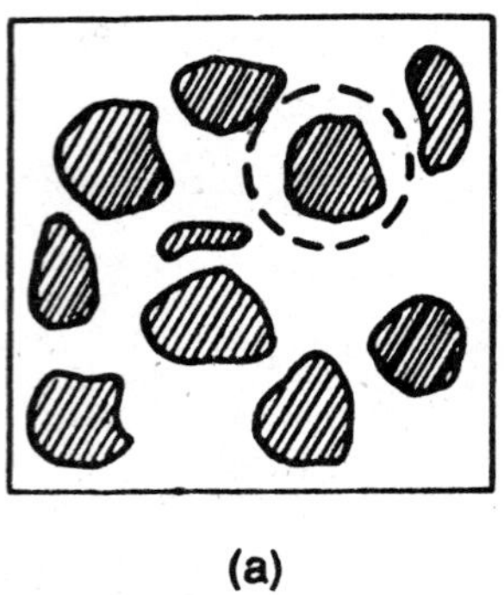

(a)

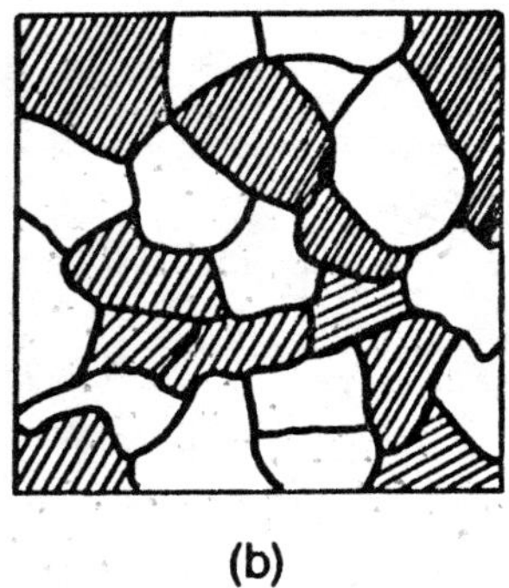

(b)

Figure 1   (a) Schematic depiction of the dispersed-inclusion microstructure. The dashed line delineates a basic structural unit for the composite.

(b) Schematic depiction of the symmetric microstructure. Depending on the relative volume fraction of the two components, either component can be the matrix phase.

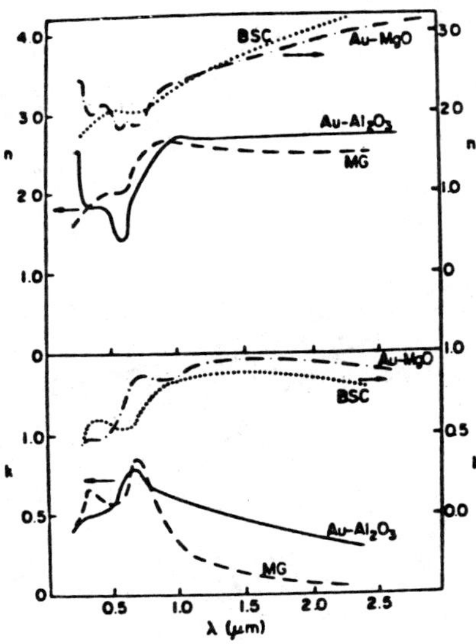

Figure 2   The real and imaginary parts of the index of refraction, n+ik, plotted
as a function of wavelength for two composite cermets. MG denotes the
Maxwell-Garnett theory, Equation (3), for dispersed inclusion microstructure, and BSG denotes the Bruggemann's self-consistent theory,
Equation (4), for the symmetric microstructure. The volume fraction of
metal in the two composites are ~30%. Figure taken from Ref. 10.

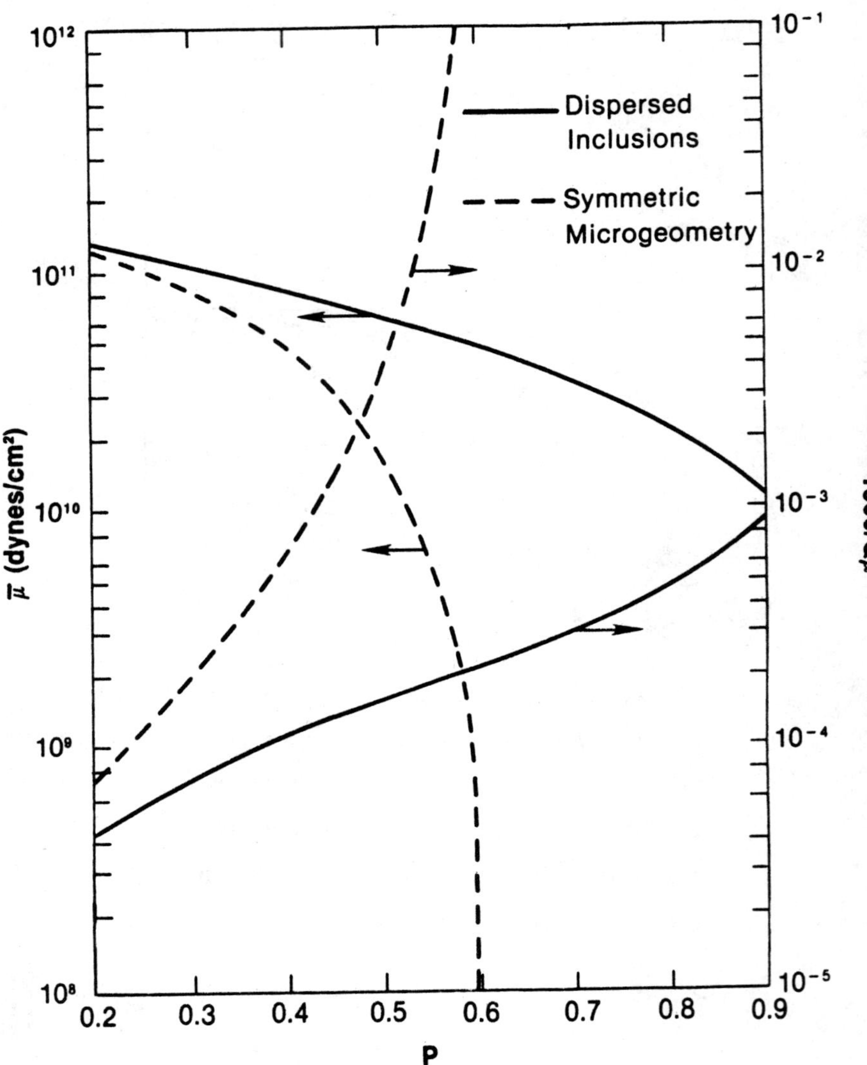

Figure 3   Calculated shear modulus and shear wave attenuation for the two composite microstructures plotted as a function of fluid concentration p.

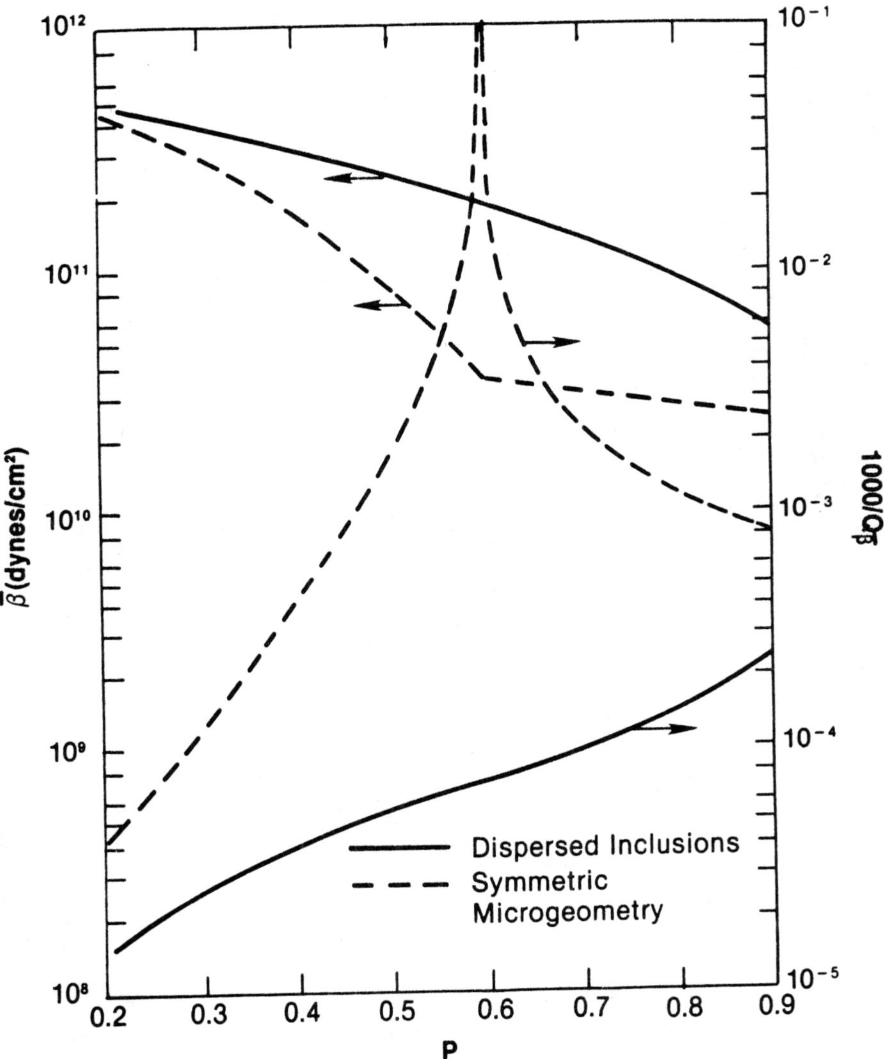

Figure 4   Calculated p-wave modulus and p-wave attenuation for the two composite microstructures plotted as a function of fluid concentration p.

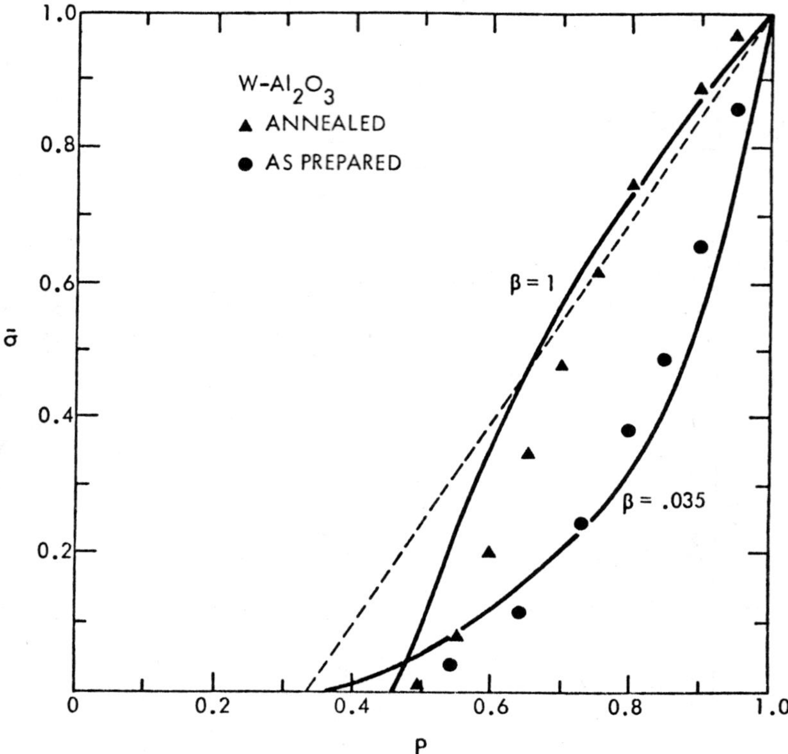

Figure 5   Normalized conductivity $\bar{\sigma}$ plotted as a function of metal volume fraction p for samples of W-Al$_2$O$_3$ cermets. The data are from B. Abeles, P. Sheng, M.D. Coutts, and Y. Arie, Adv. Phys. **24**, 407 (1975). Solid lines are calculated from theory. Dashed line denotes Bruggeman's effective medium theory prediction.

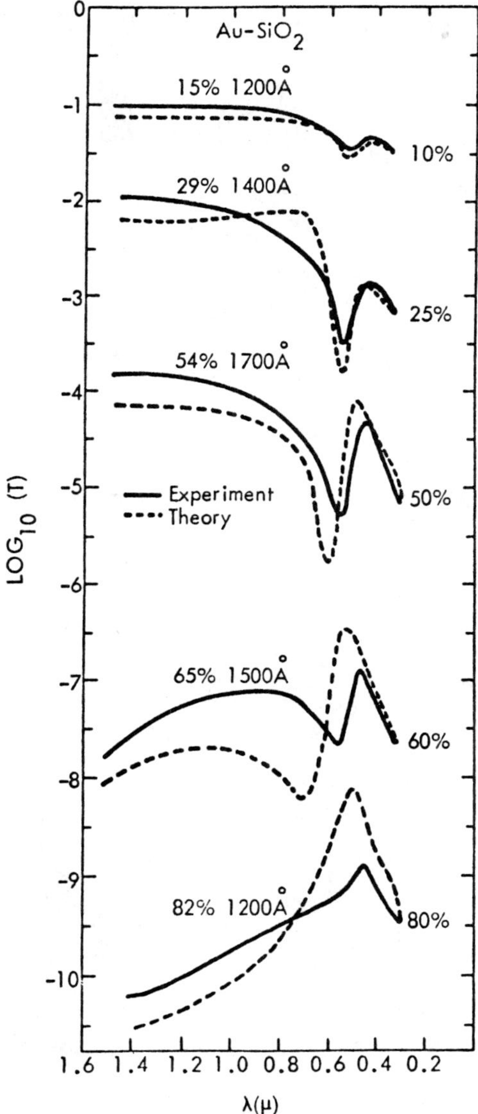

Figure 6

Optical transmission as a function of light wavelength for a series of Au-SiO$_2$ composites. Data are from R.W. Cohen, G.D. Cody, M.D. Coutts, and B. Abeles, Phys. Rev. B8, 3689 (1973). The curves are displaced with respect to one another for clarity. The theoretical curves are normalized to the experimental values at λ = 0.3 μm. The theoretical values of metal volume fraction p are labeled to the right of pairs of curves, whereas the experimental values of p and the film thick-ness are given above each pair of curves.

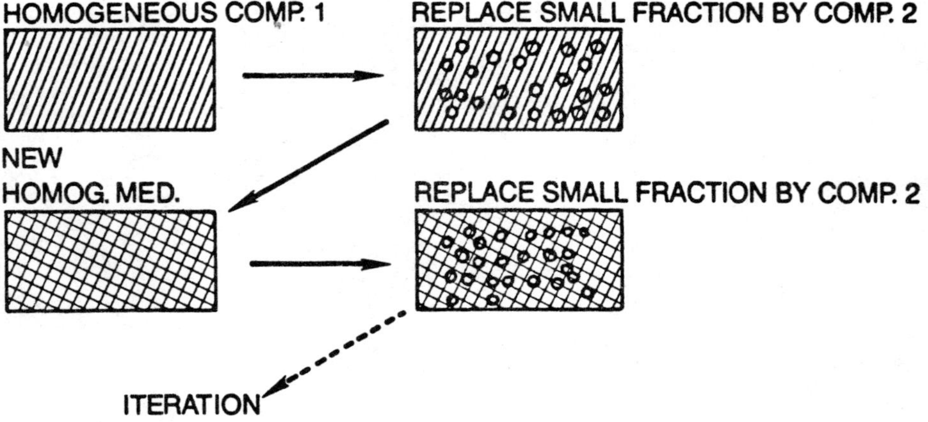

Figure 7   The basic microstructure build-up process in the DEM theory.

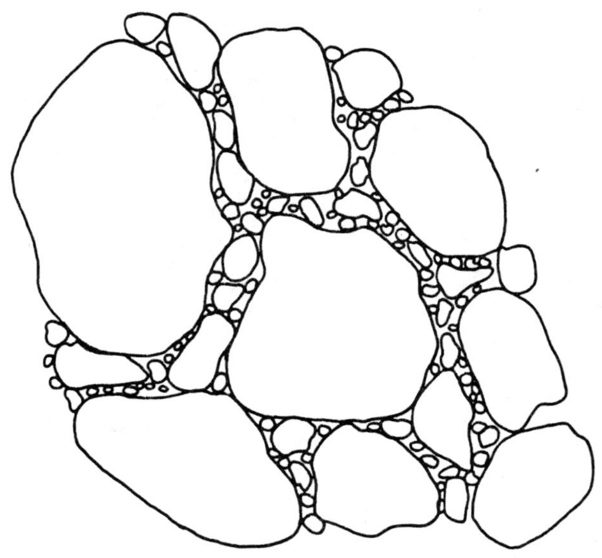

Figure 8   Schematic depiction of a DEM microstructure.

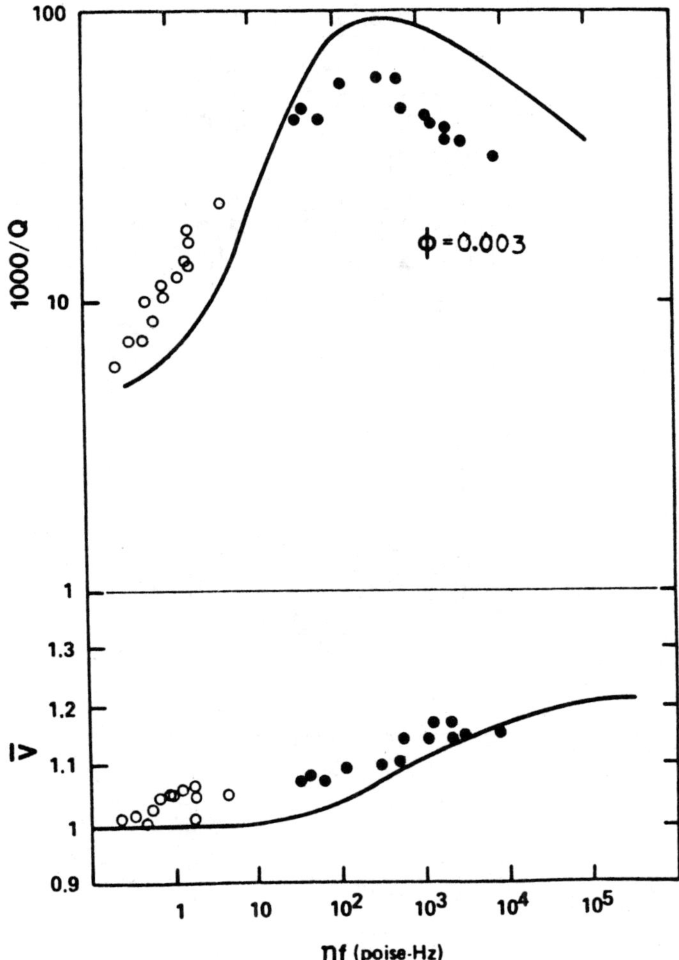

Figure 9   Attenuation per cycle, expressed in 1000/Q, and normalized acoustic velocity $\bar{V}$ (low frequency velocity = 1) plotted as a function of the product ηf (viscosity and frequency). Solid lines denote the theory, open circles denote data taken by using water as the filling fluid, and solid circles denote data taken by using glycerin as the filling fluid (data from Ref. 25).

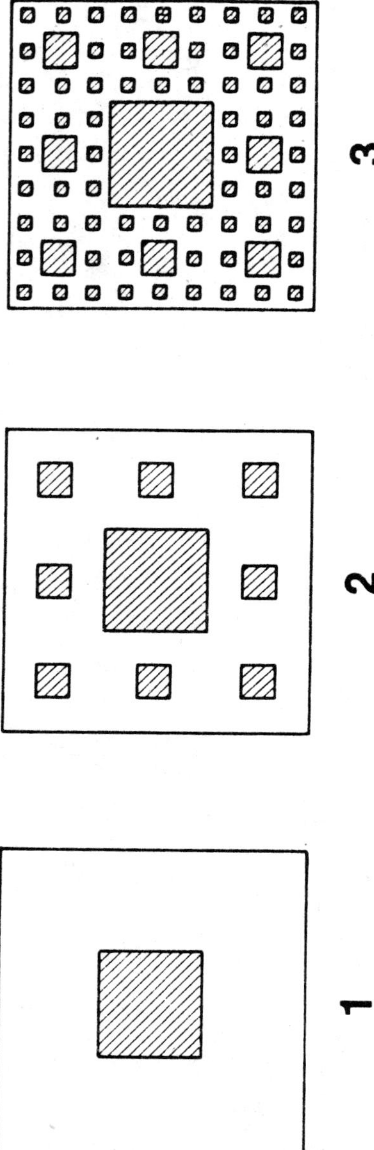

Figure 10 The three stages of a Sierpinski carpet. Area that is empty is indicated by shading. The ratio between the sides of successively sized
squares is 3.

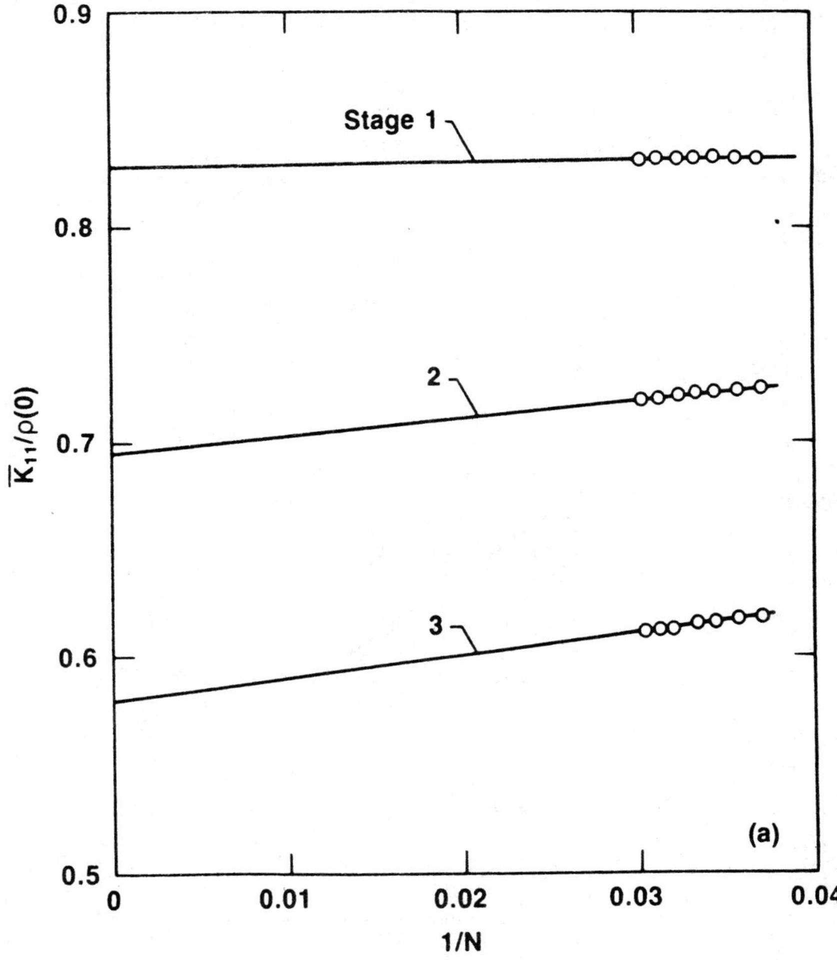

Figure 11 (a) Variation of the modulus $\overline{K}_{11}$ as a function of 1/N.

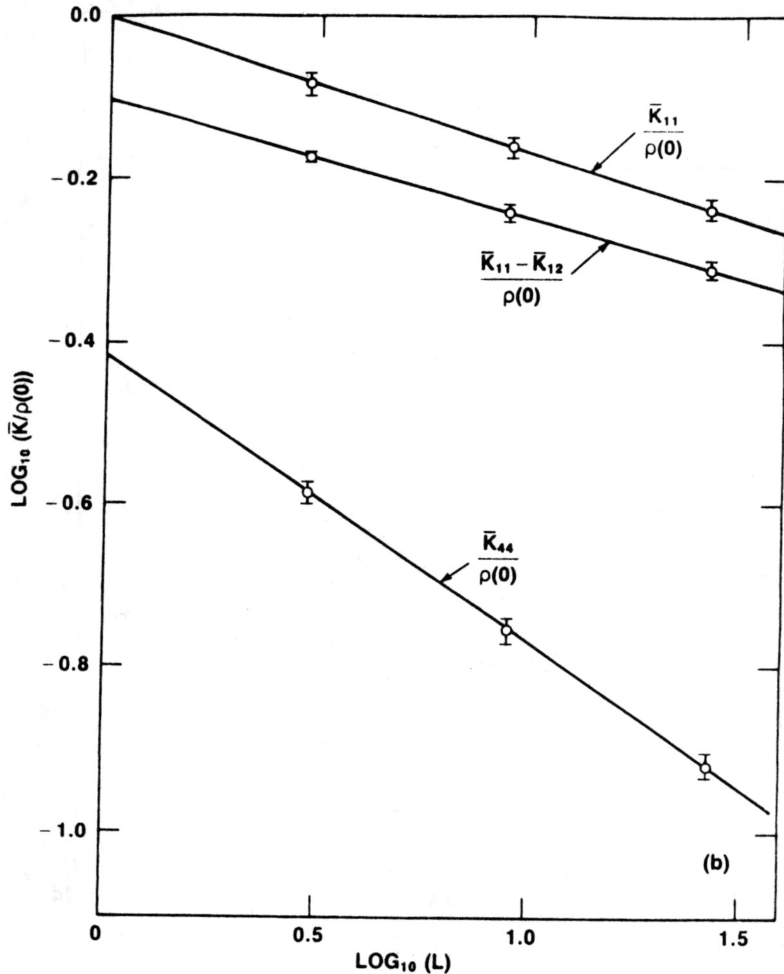

Figure 11 (b)   Variation of the carpet moduli as a function of the size parameter L.

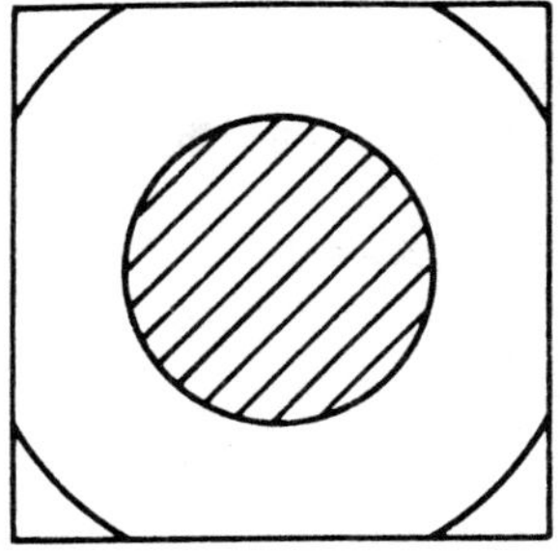

(a)

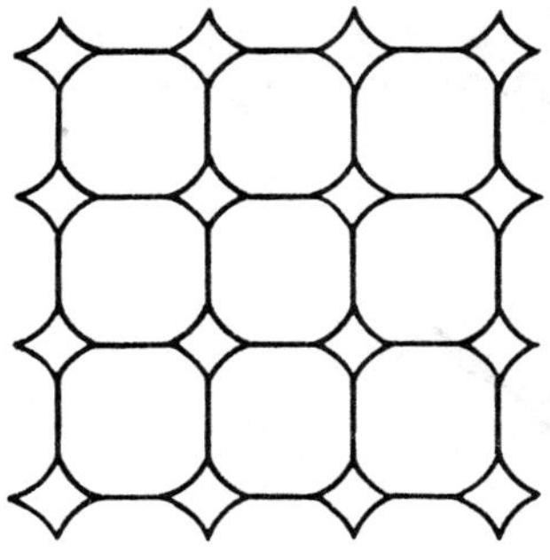

(b)

Figure 12   Geometry of the porous frame.

(a) A side view of the unit cell. The shaded area denotes the area of contact between neighboring cells (place where a cap has been cleaved off).

(b) A cross section of the porous frame.

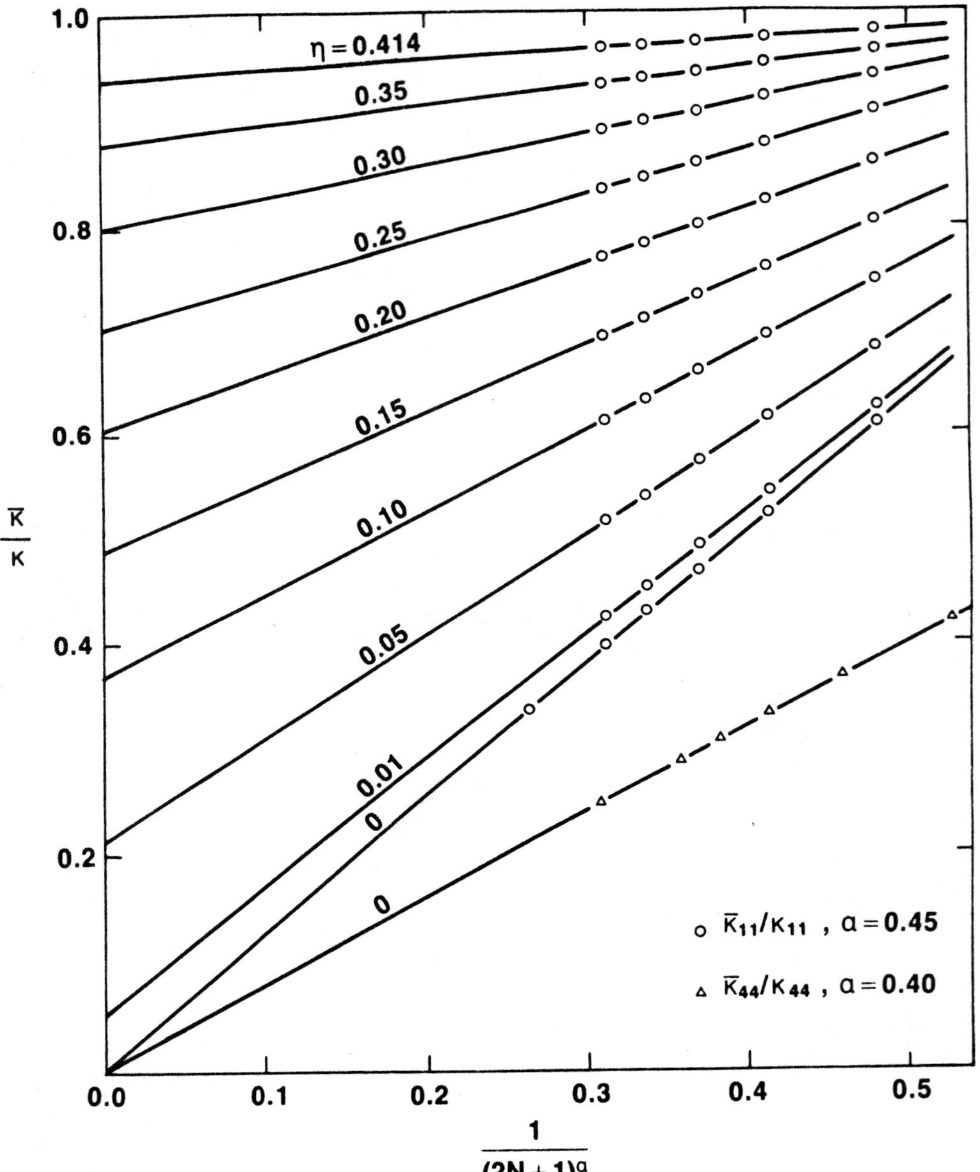

Figure 13 Variation of the calculated effective moduli with the number of reciprocal wavevectors used in the calculation. The parameters are defined in the text.

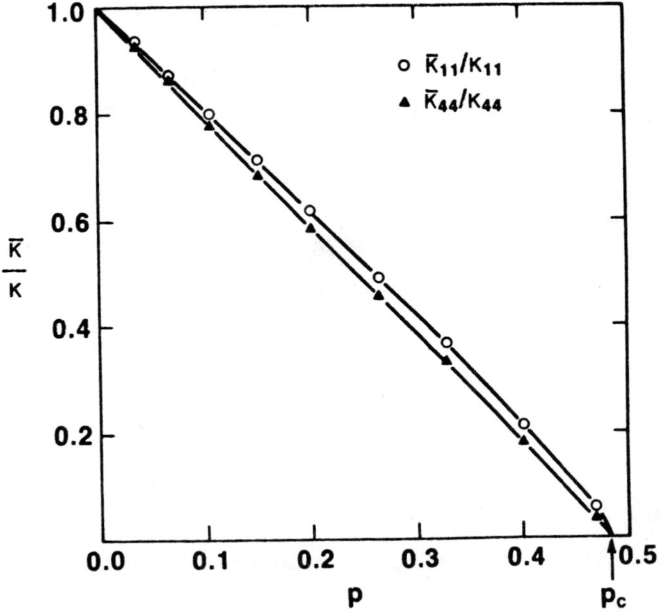

Figure 14   The normalized effective moduli plotted as a function of porosity for the porous frame.

REMARKS ON HOMOGENIZATION

Luc Tartar

Centre d'Etudes de Limeil-Valenton/M.A
B.P.27
94190 Villeneuve Saint Georges
FRANCE

Homogenization is concerned with the relations between microscopic and macroscopic scales but different mathematical problems can be associated to this general question: one of them is to give a probabilistic framework where microscopic quantitites are functions depending on a parameter ω lying in a probability space and macroscopic quantities are expectations of them [this may be the subject of another workshop], another one is to consider asymptotic expansions where one considers functions $u(x,x/\varepsilon)$, where $u(x,y)$ is periodic in y, which are called microscopic values, the macroscopic quantities being obtained by averaging in y.

The framework I have been advocating in twelve years of (partly unpublished) joint work with Francois Murat is nearer to the second one but has no periodicity assumptions: one deals with sequences u_ε which describe microscopic quantities and macroscopic quantities are limits of sequences for a suitable weak topology. In this context we will talk about oscillations of a function to describe a sequence of functions which converges weakly but may not converge strongly, homogenization is then concerned with understanding how oscillations of coefficients of a partial differential equation create oscillations in its solution.

If a problem is solved in our framework it can usually be transposed to the others (a large number of results in the second framework are applications of methods that we had derived with the purpose of being adapted to general variational situations) but some questions have not yet been defined in our framework: the use of correlations is one of them. As much as possible the theorems proved must have a local character in order to determine if the boundary conditions are crucial or have only secondary effects; of course a complete solution of a problem will then involve studying a reasonably large class of boundary conditions.

This mathematical model of some physical questions involving different scales will of course be questioned by some; it is natural that it be so but I hope that criticism will be made in a constructive way and so improve my understanding of continuum mechanics and physics (and maybe of mathematics).

I will describe now a few examples which I find typical and instructive.

I. A Problem Modeled on Stokes Equation

Ω being a bounded open set in $\mathbb{R}^3$, we consider the equation

$$\begin{cases} -\nu\Delta u_\varepsilon + u_\varepsilon \times \mathrm{curl}\ (v_0 + \lambda v_\varepsilon) = f - \mathrm{grad}\ p_\varepsilon \\ \mathrm{div}\ u_\varepsilon = 0 \end{cases} \tag{1}$$

and we assume

$$u_\varepsilon \to u_0 \quad \text{in } (H^1(\Omega))^3 \text{ weak.} \tag{2}$$

The purpose of the model is to understand how the oscillations of v_ε will create oscillations in $\mathrm{grad}\ u_\varepsilon$ and to discover the equation satisfied by u_0.

We assume that $v_0 \in (L^3(\Omega))^3$, $f \in (H^{-1}(\Omega))^3$ and that

$$v_\varepsilon \to 0 \quad \text{in } (L^3(\Omega))^3 \text{ weak.} \tag{3}$$

$\lambda > 0$ is a strength parameter which is there to emphasize the quadratic effect of the oscillations.

Although, having no time variable in it, the above model cannot describe a real situation, it is worth noticing that force terms in $u \times \mathrm{curl}\ \phi$ occur either because of a magnetic force $u \times b$ or in the Navier-Stokes equation by writing $(u \cdot \nabla)u = u \times \mathrm{curl}(-u) + \mathrm{grad}\ \frac{|u|^2}{2}$. Although this force does not work directly it creates oscillations which dissipate energy at a microscopic level, an effect that will appear in the equation satisfied by u_0 . [I have more recently written a time dependent version of this result, but it applies only to the whole space].

<u>Theorem 1</u>: There is a subsequence extracted from v_ε and a nonnegative symmetric matrix $M(x)$ constructed from this subsequence such that u_0 satisfies the equation

$$\begin{cases} -\nu\Delta u_0 + u_0 \times \mathrm{curl}\ v_0 + \lambda^2 M u_0 = f - \mathrm{grad}\ p_0 \\ \\ \mathrm{div}\ u_0 = 0 \end{cases} \tag{4}$$

and such that

$$\nu |\operatorname{grad} u_\varepsilon|^2 \to \nu |\operatorname{grad} u_0|^2 + \lambda^2 (Mu_0, u_0) \quad \text{in} \quad D'(\Omega) \qquad \Box \qquad (5)$$

<u>Remark 1</u>: The way to construct M is instructive: choose $k \in R^3$;

solve
$$\begin{cases} - \nu \Delta w_\varepsilon + k \times \operatorname{curl} v_\varepsilon = - \operatorname{grad} r_\varepsilon \\[2mm] \operatorname{div} w_\varepsilon = 0 \end{cases} \qquad (6)$$

with reasonable boundary conditions so that

$$w_\varepsilon \to 0 \quad \text{in} \quad (H^1(\Omega))^3 \quad \text{weak} . \qquad (7)$$

Then, for a subsequence, we have

$$w_\varepsilon \times \operatorname{curl} v_\varepsilon \to Mk \quad \text{in} \quad (H^{-1}(\Omega))^3 \quad \text{weak} \qquad (8)$$

$$\nu |\operatorname{grad} w_\varepsilon|^2 \to (Mk, k) \quad \text{in} \quad D'(\Omega) \qquad \Box \qquad (9)$$

<u>Proof</u>: By Sobolev's imbedding theorem $H^1(\Omega) \subset L^6(\Omega)$ (we may assume that Ω has a smooth boundary as all the results we have are local) so $u_\varepsilon \times \operatorname{curl} v_\varepsilon$ is bounded in $(H^{-1}(\Omega))^3$ and p_ε is bounded in $L^2(\Omega)/R$; we can assume that

$$u_\varepsilon \times \operatorname{curl} v_\varepsilon \to g \quad \text{in} \quad (H^{-1}(\Omega))^3 \quad \text{weak} \qquad (10)$$

$$p_\varepsilon \to p_0 \quad \text{in} \quad L^2(\Omega)/R \quad \text{weak}$$

Similarly r_ε is bounded in $L^2(\Omega)/\mathbb{R}$ and converges weakly to 0 and w_ε converges strongly to 0 in $(L^2(\Omega))^3$ by Rellich's theorem. We want to identify g and this will be done by suitable integrations by parts; notice that ⑨ follows from ⑧ by multiplying ⑥ by ϕw_ε for a test function $\phi \in \mathcal{D}(\Omega)$; symmetry of M is easily seen. Multiply ① by ϕw_ε and take limits: this gives

$$\nu \int \phi (\operatorname{grad} u_\varepsilon , \operatorname{grad} w_\varepsilon) \, dx + \lambda \int \phi (u_\varepsilon \times \operatorname{curl} v_\varepsilon , w_\varepsilon) dx \to 0 \quad \text{(note}$$

that $\int \phi (u_\varepsilon \times \operatorname{curl} v_0, w_\varepsilon) dx \to 0$ because $u_{\varepsilon j} w_{\varepsilon k} \to 0$ in $w^{1,3/2}(\Omega)$ weak) (11)

Multiply ⑥ by ϕu_ε and take limits: this gives

$$\nu \int \phi(\operatorname{grad} w_\varepsilon, \operatorname{grad} u_\varepsilon)dx + \int \phi(k \times \operatorname{curl} v_\varepsilon, u_\varepsilon)\, dx \to 0 \tag{12}$$

But because of ⑩ this gives

$$\nu \int \phi(\operatorname{grad} w_\varepsilon, \operatorname{grad} u_\varepsilon)dx \to \int \phi(g \cdot k)dx \tag{13}$$

So by ⑪ we have shown

$$\lambda \int\phi(w_\varepsilon \times \operatorname{curl} v_\varepsilon, u_\varepsilon)dx \to \int \phi(g \cdot k)dx \tag{14}$$

It remains to improve ⑧ into

$$w_\varepsilon \times \operatorname{curl} v_\varepsilon \to Mk \quad \text{in} \quad (H_{loc}^{-1}(\Omega))^3 \quad \text{strong} \tag{15}$$

to deduce that

$$g = \lambda\, MU_0 \text{ so that } u_0 \text{ satisfies } ④. \tag{16}$$

In order to prove ⑮ we notice that $w_{\varepsilon j}\dfrac{\partial}{\partial x_k} v_{\varepsilon\ell} = \dfrac{\partial}{\partial x_k}(w_{\varepsilon j}v_{\varepsilon\ell}) - v_{\varepsilon\ell}\dfrac{\partial w_{\varepsilon j}}{\partial x_k}$ and $w_{\varepsilon j}\, v_{\varepsilon\ell} \to 0$ in $L^2(\Omega)$ strong and it suffices to remark that $|\operatorname{grad} w_\varepsilon|$ is bounded in $L_{loc}^3(\Omega)$ by a regularity theorem and so that $v_{\varepsilon\ell}\dfrac{\partial w_{\varepsilon j}}{\partial x_k}$ is bounded in $L_{loc}^{3/2}(\Omega)$ and so compact in $H_{loc}^{-1}(\Omega)$. To obtain ⑤ we multiply ① by ϕu_ε which gives

$$\left\{ \begin{aligned} &\nu \int \phi\, |\operatorname{grad} u_\varepsilon|^2\, dx + \nu \sum_{i,j} \int \frac{\partial u_{0i}}{\partial x_j}\, u_{0i}\, \frac{\partial \phi}{\partial x_j}\, dx \\[2mm] &+ \int \phi\,(u_0 \times \operatorname{curl} v_0, u_0)dx \to (f, \phi u_0) + \int p_0(u_0 \cdot \operatorname{grad} \phi)dx \end{aligned} \right. \tag{17}$$

and then using ④ multiplied by ϕu_0 we obtain ⑤. $\qquad\qquad \square$

<u>Remark</u> 2: As was noticed in the proof, convergences in ⑧ and ⑨ are not optimal; using regularity theorems which give $|\operatorname{grad} w_\varepsilon|$ bounded in $L_{loc}^3(\Omega)$ we see that

$$\nu|\operatorname{grad} w_\varepsilon|^2 \to M(k,k) \quad \text{in} \quad L_{loc}^{3/2}(\Omega) \quad \text{weak} \tag{18}$$

and

$$w_\varepsilon \times \operatorname{curl} v_\varepsilon \to Mk \quad \text{in} \quad (W_{loc}^{-1,q}(\Omega))^3 \quad \text{strong for each} \quad q < 3. \tag{19}$$

so we have

232

$$M \in (L^{3/2}(\Omega))^9 \ . \qquad \qquad \qquad \square$$

<u>Remark</u> 3: It is important to notice that M does not depend on the boundary conditions imposed on 6 , as long as 7 holds. The knowledge of oscillations of v_ε in an open set ω suffices to define M on ω . If the oscillations of v_ε have a periodic structure it will be natural to impose periodicity conditions on w_ε .

In the case where $\operatorname{div} v_\varepsilon = 0$, which occurs for Navier Stokes equation 6 can be replaced by a Laplace equation

$$- \nu \ \Delta \ Z_\varepsilon = v_\varepsilon \tag{20}$$

and then

$$w_\varepsilon = (k \cdot \operatorname{grad}) Z_\varepsilon \qquad (\text{if} \ \ \operatorname{div} v_\varepsilon = 0 \ ) \tag{21}$$

so we have

$$\nu \sum_{\ell m} \frac{\partial^2 Z\ell}{\partial x_i \ \partial xm} \ \frac{\partial^2 Z\ell}{\partial x_j \ \partial xm} \to M_{i,j} \qquad (\text{in} \ \ L^{3/2}(\Omega) \ \text{weak}) \qquad \square \tag{22}$$

II. Homogenization for a diffusion equation

We now consider a situation where the oscillating coefficients are in the higher order term, the basic examples being a diffusion equation:

$$\begin{cases} - \operatorname{div} (a_\varepsilon \operatorname{grad} u_\varepsilon) = f_\varepsilon \\ u_\varepsilon \to u_0 \ \text{in} \ H^1(\Omega) \ \text{weak}, \ \Omega \ \text{open in} \ {\rm I\!R}^N \\ f_\varepsilon \to f_0 \ \text{in} \ H^{-1}(\Omega) \ \text{strong} \end{cases} \tag{23}$$

We assume that a_ε is a measurable $N{\times}N$ matrix satisfying

$$\begin{cases} (a_\varepsilon(x)\lambda,\lambda) \geqslant \alpha|\lambda|^2 \ \text{for all} \ \lambda \ \ R^N \ \text{a.e.} \ x \ \ \Omega \\ \\ (a_\varepsilon^{-1}(x)\lambda,\lambda) \geqslant \beta^{-1}|\lambda|^2 \ \text{for all} \ \lambda \ \ R^N \ \text{a.e.} \ x \ \ \Omega \\ \\ \text{with} \ 0 < \alpha \leqslant \beta < +\infty \end{cases} \tag{24}$$

In applications a_ε is symmetric and (25) can be transformed into

$$\alpha|\lambda|^2 < (a_\varepsilon(x)\lambda,\lambda) < \beta|\lambda|^2 \quad \text{for all} \quad \lambda \in \mathbb{R}^N \quad \text{a.e.} \quad x \in \Omega \qquad (24')$$

Then we have the following theorem

Theorem 2: If a sequence a_ε satisfies (24), then there is a subsequence and an a_0 satisfying inequalities (24) with the same constants α,β, such that: if a sequence u_ε satisfies (23) then this implies

$$a_\varepsilon \text{grad } u_\varepsilon \to a_0 \text{grad } u_0 \quad \text{in} \quad L^2(\Omega)^N \quad \text{weak} \qquad (25)$$

If each a_ε is symmetric, then a_0 is symmetric $\qquad\qquad \square$

The proof relies on the construction of test functions satisfying an adjoint equation

$$- \text{div } (a_\varepsilon^* \text{grad } v_\varepsilon) = q_\varepsilon \qquad (26)$$

which, thanks to (24), can be solved in a functional analytic way; then multiply (23) by ϕv_ε, (26) by ϕu_ε with $\phi \in D(\Omega)$ and integrate by parts: this gives the results if q_ε has been choosen correctly and a suitable subsequence has been extracted.

It is useful to understand the integration by parts performed here as a simple example of a compensated compactness argument:

Lemma 1: If
$$\begin{cases} E_\varepsilon \to E_0 \quad \text{in} \quad L^2(\Omega)^N \quad \text{weak} & (27) \\[2ex] \dfrac{\partial E_{\varepsilon i}}{\partial x_j} - \dfrac{\partial E_{\varepsilon j}}{\partial x_i} \quad \text{compact of} \quad H^{-1}_{loc}(\Omega), \text{ for each } i,j. \end{cases}$$

and
$$\begin{cases} D_\varepsilon \to D_0 \quad \text{in} \quad L^2(\Omega)^N \quad \text{weak} & (28) \\[2ex] \displaystyle\sum_j \dfrac{\partial D_{\varepsilon j}}{\partial x_j} \in \text{compact of} \quad H^{-1}_{loc}(\Omega) \end{cases}$$

then we have

$$\sum_j E_{\varepsilon j} D_{\varepsilon j} \to \sum_j E_{0j} D_{0j} \quad \text{in} \quad D'(\Omega) \qquad\qquad \square \quad (29)$$

We can then remark that there is a new type of weak topology on a_ε and we will denote $a_\varepsilon \overset{H}{\to} a_0$ to say that ⓐ⑦, ⓐ⑧ and

$$D_\varepsilon = a_\varepsilon E_\varepsilon \qquad (30)$$

implies

$$D_0 = a_0 E_0 . \qquad (31)$$

Theorem 2 says that the set of a_ε defined by inequalities 24 for some α, β , is compact for this H-convergence.

Remark 4: On a set defined by ㉔ this topology is metrizable, and in the symmetric case coincides with the G-convergence studied by DeGiorgi, Marino and Spagnolo; the above setting has the advantage of showing more clearly the local character of the convergence, the role of lemma 1, and, not being related to any minimization of a functional, of handling more general situations of continuum mechanics or physics. ☐

Remark 5: A good way to understand the nature of this convergence is to use an electrostatic interpretation for equation ㉓: the important physical quantitities u, E, D, ρ, a, e are related by

$$\begin{cases} a) & E_\varepsilon = - \text{grad } u_\varepsilon \\[4pt] b) & \text{div } D_\varepsilon = \rho_\varepsilon \\[4pt] c) & D_\varepsilon = a_\varepsilon E_\varepsilon \\[4pt] d) & e_\varepsilon = \frac{1}{2} E_\varepsilon \cdot D_\varepsilon \end{cases} \qquad (32)$$

Equations a) b) are linear and are conserved by weak convergence; equation d) is non linear but if ρ_ε stays in a compact of $H^{-1}_{loc}(\Omega)$ and using curl $E_\varepsilon = 0$ we can apply lemma 1.

Handling c) requires the right convergence on a_ε . In other terms u, E, D, ρ, e define differential forms of order 0,1,N-1, N,N; equation a) and b) involve exterior derivative and d) involve exterior product: weak topology is natural for coefficients of differential forms. Equation c) explains that a transforms 1-form into N-1 form and, with information a) b) this defines the right topology to put on a_ε . ☐

An important result for understanding H-convergence is the formula for laminates, that is the case where the coefficients oscillate only in one direction: assume for example that a^ε is a function of x_1 alone. The result is given by the

__Lemma__ $\underline{2}$: If a^ε only depends on x_1, then $a^\varepsilon \xrightarrow{H} a^0$ means

a) $\dfrac{1}{a^\varepsilon_{11}} \to \dfrac{1}{a^0_{11}}$ in L^∞ weak $*$

b) $\dfrac{a^\varepsilon_{i1}}{a^\varepsilon_{11}} \to \dfrac{a^0_{i1}}{a^0_{11}}$ in L^∞ weak$*$ for $i \neq 1$

b)' $\dfrac{a^\varepsilon_{1j}}{a^\varepsilon_{11}} \to \dfrac{a^0_{1j}}{a^0_{11}}$ in $L\infty$ weak $*$ for $j \neq 1$

c) $a^\varepsilon_{ij} - \dfrac{a^\varepsilon_{i1} a^\varepsilon_{ij}}{a^\varepsilon_{11}} \to a^0_{ij} - \dfrac{a^0_{i1} a^0_{1j}}{a^0_{11}}$ in L^∞ weak $*$ for $i \neq 1$, $j \neq 1$ $\square$

These, apparently strange, quantities appear in a natural way if instead of expressing D_ε in terms of E_ε, one expresses $(E_{1\varepsilon}, D_{2\varepsilon}, \ldots D_{N\varepsilon})$ in term of $(D_{1\varepsilon}, E_{2\varepsilon}, \ldots E_{N\varepsilon})$ and this is a simple case of a more general procedure which can be followed for more general, linear or nonlinear, equations in the laminate case. The quantities $D_1, E_2, \ldots E_N$ are the only one which can be defined as traces on hyperplanes $x_1 =$ constant under the hypothesis div $D \in L^2(\Omega)$ and curl $E \in L^2(\Omega)^{N^2}$; in an other way they are defined by the restriction to the hyperplane of the differential forms E and D.

The procedure consists in looking, among the quantities satisfying differential equations with constant coefficients, for those who have a good derivative in x_1: more precisely if $v_{i\varepsilon} \to v_{i0}$ in $L^2(\Omega)$ weak, $\Sigma \dfrac{\partial}{\partial x_j} v_{i\varepsilon} \in$ compact of $H^{-1}_{loc}(\Omega)$ and $v_{i\varepsilon}$ bounded in $L^2(\Omega)$, then $v_{1\varepsilon} g^\varepsilon(x_1) \to v_{10} g^0(x_1)$ if $g^\varepsilon \to g^0$ in L^2 weak, by a simple application of Lemma 1. If we have found enough of these quantities in order to express the others, then we have a constitutive relation of the form $w^\varepsilon = b^\varepsilon(x_1) v^\varepsilon$ which gives $w^0 = b^\circ(x_1) v^\circ$ for the weak limits. On the other hand, this also gives a particular solution with v constant and $w^\varepsilon = b^\varepsilon(x_1) v$.

If we apply this procedure to a linear elasticity problem:

$\sigma_{ij}^{\varepsilon} = c_{ijkl}^{\varepsilon}(x_1)\, e_{kl}^{\varepsilon}$ with $e_{kl}^{\varepsilon} = \frac{1}{2}\left(\frac{\partial \hat{u}_k^{\varepsilon}}{\partial x_1} + \frac{\partial \hat{u}_1^{\varepsilon}}{\partial x_k}\right)$, we first notice that from the

equilibrium equation $\sum_j \frac{\partial}{\partial x_j}\, \sigma_{ij}^{\varepsilon} = f_i^{\varepsilon}$ that the $\sigma_{i1}^{\varepsilon}$ do not oscillate (and by sym-

metry the $\sigma_{1i}^{\varepsilon}$) : for $k\neq 1$, $\ell\neq 1$ e_{i1}^{ε} does not oscillate as neither $\frac{\partial u_k^{\varepsilon}}{\partial x_{\ell}}$ nor

$\frac{\partial u_{\ell}^{\varepsilon}}{\partial x_k}$ do. We should then express everything in terms of the σ_{i1} and the

ε_{kl} ($k\neq 1$, $\ell\neq 1$); a matrix $K^{\varepsilon}(x_1)$ will then appear and its limit should be

taken. If we restrict to isotropic materials

$$\sigma_{ij} = 2\mu(x_1)e_{ij} + \lambda(x_1)\delta_{ij}\sum_k e_{kk} \tag{34}$$

we should write (we work in dimension N arbitrary)

$$\begin{cases}
e_{11} = \dfrac{1}{2\mu+\lambda}\,\sigma_{11} - \dfrac{\lambda}{2\mu+\lambda}\left(\sum_{k\neq 1} e_{kk}\right) \\[2ex]
\sigma_{ii} = \dfrac{\lambda}{2\mu+\lambda}\,\sigma_{11} + 2\mu\, e_{ii} + \dfrac{2\mu\lambda}{2\mu+\lambda}\left(\sum_{k\neq 1} e_{kk}\right) \quad \text{if } i\neq 1 \\[2ex]
\sigma_{ij} = 2\mu\, e_{ij} \quad \text{if } i\neq 1,\ j\neq 1 \\[2ex]
e_{i1} = \dfrac{1}{2\mu}\,\sigma_{i1} \quad \text{if } i\neq 1
\end{cases} \tag{35}$$

The formula for laminates then consists in replacing the 5 different quantities

$\frac{1}{2\mu+\lambda}$, $\frac{\lambda}{2\mu+\lambda}$, $\frac{\lambda\mu}{2\mu+\lambda}$, μ, $\frac{1}{\mu}$ by their weak limits. The algebraic condition

that ensures that such relations can be found is that the matrix of entries

$M_{ik} = C_{i1k1}$ be invertible, a consequence of strong ellipticity:

$\sum_{ijkl} C_{ijkl}\,\xi_i n_j \xi_k n_1 \geq \alpha |\xi|^2 |n|^2$.

Remark 6: In order to prove an homogenization theorem for linear elasticity and

general coefficients, the construction of test functions relies on the

Lax-Milgram lemma which requires the very strong ellipticity condition:

$\sum_{ijkl} C_{ijkl}\, e_{ij} e_{kl} \geq \alpha |e|^2$. I do not know if this can be avoided. In the laminate

case this difficulty disappears.

In the nonlinear case the above analysis provides the construction of test

functions: in nonlinear elasticity it requires a strict Legendre Hadamard con-

dition with growth conditions; but I do not know under what kind of reasonable

hypothesis an homogenization theorem holds. $\qquad\square$

Remark 7: In some situations, like plate problems, the formula for a laminate follows from the following generalization:

If $U^\varepsilon_{ij} \to U^0_{ij}$ in $L^2(\Omega)$ weak and $\sum\limits_{i,j} \dfrac{\partial^2 U^\varepsilon_{ij}}{\partial x_i \partial x_j}$ compact of $H^{-2}_{loc}(\Omega)$ then

$U^\varepsilon_{11} f^\varepsilon(x_1) \to u^0_{11} f^\circ(x_1)$ if $f^\varepsilon \to f^\circ$ weakly. Homogenizing an equation $\sum\limits_{ijkl} \dfrac{\partial^2}{\partial x_i \partial x_j}$

$A^\varepsilon_{ijkl}(x_1) \dfrac{\partial^2}{\partial x_k \partial x_l}$ then requires taking limits of

$$\frac{1}{A^\varepsilon_{1111}} \; ; \; \frac{A^\varepsilon_{11kl}}{A^\varepsilon_{1111}} \quad \text{for} \quad k \text{ or } l \neq 1 \qquad A^\varepsilon_{ijkl} - \frac{A^\varepsilon_{ij11} A^\varepsilon_{11kl}}{A^\varepsilon_{1111}} \quad \text{for } (i,j) \neq (1,1)$$

and $(k,l) \neq (1,1)$　　　　　　　　　□

Remark 8: Understanding which components of the different fields oscillate and which do not is an important step and this is certainly useful in designing experiments: if you try to "measure" an oscillating quantity you will either get an average result, that may be different from the one you want, or (and this clearly happens now when measuring probes become smaller) get a pointwise result from which the information is certainly hard to extract (and you may think that it is not your fault because you are facing a chaotic process!).

I believe that most of the strange rules guessed by physicists are just an attempt to understand some kind of homogenization process (in a more difficult framework of oscillations of semilinear hyperbolic systems); I hope to make this idea more precise in the future.　　　　　　　　□

Once the formula for laminates is understood, the next step is to get rid of indices and write the same formula in a more intrinsic, and useful, way.

Assume that we mix two different materials of matrix A,B with proportion θ and $(1-\theta)$ in slices perpendicular to a vector e; then we get an effective matrix C given by one of the two equivalent formulas

$$\begin{cases} (C-A)^{-1} = \dfrac{(B-A)^{-1}}{1-\theta} + \dfrac{\theta}{1-\theta} \; \dfrac{e \times e}{(Ae,e)} \\[3ex] (C-B)^{-1} = \dfrac{(A-B)^{-1}}{\theta} + \dfrac{1-\theta}{\theta} \; \dfrac{e \times e}{(Be,e)} \end{cases} \tag{36}$$

To see this we consider a solution such that grad u = a when material has matrix A and grad u = b when it has matrix B;

The vector $E(x)$ = a or b will then be a gradient if the following compatibility condition holds:

$$b = a + te \text{ for some } t \in \mathbb{R} .\tag{37}$$

The vector $D(x)$ = Aa or Bb will then be divergence free if the following compatibility condition holds:

$$(Bb - Aa \cdot e) = 0 \tag{38}$$

Then (37) and (38) give t as a function of a:

$$t(Be,e) + ((B-A)a \cdot e) = 0 \tag{39}$$

The average fields E and D are given by

$$\begin{cases} E = \theta a + (1-\theta)b \\ D = \theta Aa - (1-\theta)Bb \end{cases} \tag{40}$$

The matrix C is defined by

$$D = CE \tag{41}$$

A simple algebraic manipulation will then give formula 36) for C $\square$

The next step is then to use formula (36) to reiterate the process: having obtained a material of matrix C_{n-1} we then mix it with the material of matrix B in slices perpendicular to e_n and this gives a material of matrix C_n .

A simple calculation then shows that mixing A and B with proposition θ and $(1-\theta)$ can lead to a material of matrix C given by

$$(C-B)^{-1} = \frac{(A-B)^{-1}}{\theta} + \frac{1-\theta}{\theta} \sum_j n_j \frac{e_j \otimes e_j}{(Be_j,e_j)} \tag{42}$$

$$\text{with } n_j > 0 , \sum_j n_j = 1$$

Similar constructions can be done for more general equations with more tedious calculations (Francois Murat and Gilles Francfort have carried it out for linear elasticity.)

The formula (42) is then easy to use to construct a large class of materials using $A = \alpha I$ and $B = \beta I$ with proportion θ and $1-\theta$. Showing that this class is optimal is less obvious; but let us state the result first:

<u>Theorem</u> <u>3</u>: Let χ_ϵ be a sequence of characteristic functions converging weakly to θ and assume that $a_\epsilon = \chi_\epsilon \alpha I + (1-\chi_\epsilon) \beta I$ H-converges to $C(x)$. Then the eigenvalues $\lambda_1,\ldots,\lambda_n$ of $C(x)$ satisfy

$$\sum_j \frac{1}{\lambda_j - \alpha} < \frac{1}{\mu_-(\theta) - \alpha} + \frac{N-1}{\mu_+(\theta) - \alpha} \, a.e.$$

$$\sum_j \frac{1}{\beta - \lambda_j} < \frac{1}{\beta - \mu_-(\theta)} + \frac{N-1}{\beta - \mu_+(\theta)} \quad a.e. \tag{43}$$

$$\mu_-(\theta) < \lambda_j < \mu_+(\theta) \; \forall_j \quad a.e.$$

where we have noted

$$\mu_-(\theta) = (\frac{\theta}{\alpha} + \frac{1-\theta}{\beta})^{-1} \; ; \; \mu_+(\theta) = \theta\alpha + (1-\theta)\beta \, . \tag{44}$$

Conversely if C has its eigenvalues satisfying (43) one can find a sequence χ_ϵ . $\square$

Although this theorem gives an optimal answer to the special situation of mixing two isotropic materials for a diffusion equation, the general question is far from being understood; even for a simple equation of diffusion like (23) the result of mixing three isotropic materials or of mixing even one anisotropic material has not been done (this means that $a_\epsilon = R_\epsilon^* AR$ where A is a fixed symmetric matrix and $R_\epsilon(x)$ is an orthogonal matrix depending measurably in x), at least in a form like theorem 3: G. Milton has discussed in his talk some improvements he has made on earlier results by Schulgasser.

The method for obtaining necessary conditions is also not very satisfactory; it is based on a compensated compactness argument which seems general but has to be improved.

In the process of the proof of theorem 2, test functions are constructed which give two sequences of matrices satisfying:

$$P_\epsilon \to I \text{ in } L^2(\Omega)^{N^2} \text{ weak } ; \; Q_\epsilon = A_\epsilon P_\epsilon \to A_0 \text{ in } L^2(\Omega)^{N^2} \text{ weak}$$
$$\text{curl } (P_\epsilon \lambda) \in \text{compact } H_{loc}^{-1}(\Omega)N^3 \text{ and} \tag{45}$$
$$\text{div } (Q_\epsilon \lambda) \in \text{compact } H_{loc}^{-1}(\Omega) \text{ for each } \lambda \in R^N$$

The idea is to use the constraints on the oscillations of P_ε and Q_ε imposed by the differential equations (and understanding what they are is the goal of compensated compactness) and compare to the constraints imposed by the relation $Q_\varepsilon = a_\varepsilon P_\varepsilon$ from what is known about the sequence A_ε .

Let us choose a function $F(P,Q)$ which has the following property:

$$\begin{cases} \text{If } P_\varepsilon \to P_0 \,,\, Q_\varepsilon \to Q_0 \quad \text{in } L^2(\Omega)N^2 \quad \text{weak} \\ \text{and} \quad \mathrm{curl}\,(P_\varepsilon \lambda) \in \text{compact } H^{-1}_{loc}(\Omega)^{N^2} \\ \qquad\qquad \mathrm{div}(Q_\varepsilon \lambda) \in \text{compact } H^{-1}_{loc}(\Omega), \text{ for each } \lambda \ R^N \qquad (46) \\ \text{then for each } \phi \in \infty(\Omega), \phi \geqslant 0 \quad \liminf \int \phi F(P_\varepsilon, Q_\varepsilon) \, dx \geqslant \int \phi F(P_0, Q_0) dx \end{cases}$$

Then define for a matrix a

$$g(a) = \mathop{\mathrm{Sup}}_{P} F(P, a\,P) \qquad\qquad (47)$$

then we have the

Lemma 3: If $A_\varepsilon \overset{H}{\to} A_0$ and $\widehat{48}$ $q(a^\varepsilon) \to h$ in $L^\infty(\Omega)$ weak $*$ then $\widehat{49}$ $g(a^0) \leqslant h$. a.e. $\qquad\qquad \square$

What are the functions F satisfying $\widehat{46}$ is still an open question, but the basic result of compensated compactness characterizes the quadratic ones; unfortunately it does not tell how to construct such functions but how to check if a quadratic one is right. This is indeed the weak point of the method presented here; in our situation we have in mind to mix isotropic materials αI and βI and it is then natural to use a function F that will stay invariant under an orthogonal transformation. If we then look only at quadratic functions using $\mathrm{tr}(P^*\,P)$, $\mathrm{tr}\,(Q^*\,Q)$, $\mathrm{tr}\,(P^*\,Q)$, $\mathrm{tr}\,P$, $\mathrm{tr}\,Q$, it does not take too long to select the special functions F that we have used to prove theorem 3.

In the case where a_0 is isotropic, $\widehat{43}$ gives the bounds that were derived by Hashin and Shtrikman, their arguments have been made mathematically rigorous since and J.R. Willis has discussed in his talk of a nonlinear situation using similar ideas.

There have been some, partially successful, attempts to prove theorem 3 using other ideas (R. Kohn mentioned to me some of his); as ours, they suffer, I believe, from not being general enough. I have some hope in using D. Bergman's ideas (and still compensated compactness) to make a step towards understanding more general situations.

Understanding homogenization is not only proving theorems like our theorem 2, but going forward and proving results like theorem 3. The subject, from that point of view, is far from being closed.

III. A problem of optimal shape.

The reason why, unaware of what had been done before, I was discovering homogenization and questions of optimal bounds with F. Murat twelve years ago, was a problem of optimal shape. Realistic problems of this type have appeared since and were discussed in the talks of M. Bendsoe and of R. Kohn.

Consider a bounded open domain Ω of R^N ; the unknown is a measurable subset ω constrained by

$$\text{meas } \omega = \gamma \tag{50}$$

The subset ω is filled with a material αI and its complement with a material βI , so we have a function c taking only values α or β and a matrix $a = cI$. The state of our system is solution of

$$\begin{cases} -\text{div (a grad u)} = f \\ u\big|_{\partial\Omega} = 0 \end{cases} \tag{51}$$

and the function to maximize is given by

$$J_0(\omega) = \int_\omega q(x,u)dx + \int_{\Omega\setminus\omega} h(x,u)dx \tag{52}$$

In many instances this kind of problem has no optimal solution and one has to introduce a relaxed problem: homogenization appears because a mixture of the two materials may give a better value than any subset ω . A first relaxed problem can then be:

$$\text{Maximize } J_1(\theta,a_0) = \int_\Omega(\theta q(x,u) + (1-\theta)h(x,u))dx \tag{53}$$

where u is the solution of

$$\begin{cases} -\text{div } (a_0 \text{gradu}) = f \\ u|_{\partial\Omega} = 0 \end{cases} \qquad (54)$$

and the constraints on (θ, a_0) are

$$0 < \theta_0 < 1.\text{a.e.} \qquad \int_\Omega \theta(x)dx = \gamma \qquad (55)$$

and ⑤⑥ $a_0(x)$ has its eigenvalues satisfying ㊽ with $\theta(x)$, a.e.

This new problem has a solution and necessary condition of optimality can easily be derived (for 0 given, ㊽ defines a convex set of symmetric matrices); from these optimality conditions it can be shown that a solution exists which (a.e.) has the matrix a_0 corresponding to a laminate (i.e. has one eigenvalue $\mu_-(0)$ and $(N-1)$ eigenvalues $\mu_+(0)$). A simpler problem can then usually be written.

For example if one wants to maximize

$$J_1(\omega) = \int_\Omega f(x)\, u(x)\, dx \qquad (\text{same } f \text{ than in } \text{㉛}) \qquad (56)$$

we have the relaxed problem of maximizing the same function but where u is solution of

$$\begin{cases} -\text{div } (\mu_-(\theta)\text{grad } u) = f \\ u|_{\partial\Omega} = 0 \end{cases} \qquad (57)$$

and θ satisfies ⑤⑤.

If instead one wants to minimize J_1, u must then be solution of

$$\begin{cases} -\text{div } (\mu_+(\theta)\text{grad } u) = f \\ u|_{\partial\Omega} = 0 \end{cases} \qquad (58)$$

The characterization of all possible mixtures appears as a technical step because at the beginning it was not clear what kind of mixture the optimal solution tried to produce; but as soon as this matter is understood one realizes that the complete characterization of theorem 2 is not necessary.

It is hoped that these techniques will be improved so one could attack more general problems of this type but also others like instabilities of interfaces or the structure of complex flows.

IV. A problem with a non local effect.

Homogenization problems sometimes lead to a change of type of the equation. It is actually striking that the Green's function for equation ㉓ converges to a Green's function of a similar problem; one could expect more general problems, like pseudo-differential operators: it is not difficult, and it was done by J.L. Lions, to produce such an example using a Laplace transform. E. Sanchez-Palencia studied realistic situations of this type like viscoelastic behaviour but more has to be done in this direction: for example what kind of memory effect can appear (the following analysis will use some ideas that I had tried for this characterization).

The following example is purely academic and was studied with F. Murat as a case with a lack of compactness where ideas used for theorem 2 will fail. As a result of our analysis, they indeed fail, because the limiting problem has an integral term; it is a non local effect but of a different nature: the convolution is in space, not time. Our problem has a spatial coherence but rapid fluctuations in time of the coefficients induce the strange effect of a nonlocal term in space [we are interested here by describing a weak limit of the solutions; if we were measuring a pointwise value we may experience a "chaotic" response!].

We consider the following problem

$$\begin{cases} - a_\varepsilon(t) \dfrac{d^2 u_\varepsilon}{dx^2} + b_\varepsilon(t)\, u_\varepsilon = f(x) \\[2mm] u_\varepsilon(t) \in H^1(\mathbb{R}) \ . \quad f \in L^2(\mathbb{R}) \quad t \in [0,t] \end{cases} \tag{59}$$

where the coefficients $a_\varepsilon(t)$, $b_\varepsilon(t)$ satisfy

$$\begin{cases} 0 < \alpha \leqslant a_\varepsilon(t) \leqslant \beta \\[2mm] 0 < \alpha' \leqslant \dfrac{b_\varepsilon(t)}{a_\varepsilon(t)} \leqslant \beta' \end{cases} \tag{60}$$

We assume that

$$\frac{1}{a_\varepsilon(t)} \to \frac{1}{a_0(t)} \quad \text{in } L^\infty(0,T) \text{ weak } *$$

$$\tag{61}$$

$$\frac{b_\varepsilon^m(t)}{a_\varepsilon^{m+1}(t)} \to \frac{b_m(t)}{a_0(t)^{m+1}} \quad \text{in } L^\infty(0,T) \text{ weak } * \text{ for } m \geqslant 1.$$

we then have the

Theorem __4__: $u_\varepsilon \to u_0$ in $L^2 \times (R_x(0,T))$ weak where u_0 satisfies

$$\begin{cases} u_0 \in H^1(\mathbb{R}) \quad \text{a.e.t.} \\ -a_0(t)\dfrac{d^2 u_0}{dx^2} + B_1(t)u_0 - \int_{\mathbb{R}} H(x-y,t)\, u_0(y)dy = f(x). \end{cases} \tag{62}$$

and $H(x,t)$ can be determined from $A_0, B_1, \dots B_m \dots$. $\quad\square$

To see this we will perform a Fourier transform in x and note
$$\hat{f}(\xi) = \int_R f(x)e^{-2i\pi x\xi}\, dx \; ; \quad \text{(59)} \quad \text{then becomes}$$

$$\hat{u}_\varepsilon(\xi,t) = \hat{f}(\xi)[4\pi^2|\xi|^2 a_\varepsilon(t) + b_\varepsilon(t)]^{-1} \tag{63}$$

In order to study the limit of this quantity we will use the Young's measures associated to the sequence $(a_\varepsilon, b_\varepsilon)$: there exists a probability measure $d\nu_t$ in the variables (λ_1,λ_2) whose support, because of hypothesis (60), lies in $\{\alpha < \lambda_1 < \beta \; ; \; \alpha' < \dfrac{\lambda_2}{\lambda_1} < \beta'\}$ such that

$$\frac{1}{z\, a_\varepsilon(t)+b_\varepsilon(t)} \to \phi_t(\xi) = \int \frac{d\nu_t}{z\lambda_1 + \lambda_2} \quad \text{in} \quad L^\infty(0,T) \text{ weak } * \tag{64}$$

But from (61) one derives easily
$$\begin{cases} \phi_t(z) = \dfrac{1}{a_0(t)z} - \dfrac{B_1(t)}{(A_0(t)z)^2} + \dfrac{B_2(t)}{(a_0(t)z)^3} - \cdots \\[2mm] \text{for} \quad z > -\alpha' \end{cases} \tag{65}$$

But the more important property, consequence of (64) and (60) is that ϕ can be defined for complex z outside the segment $[-\beta',\alpha']$ on the real axis and:
$$\text{Im}\phi(z) > 0 \quad \text{if} \quad \text{Im} z > 0 \tag{66}$$

Zeroes and poles of ϕ are on the segment $[-\beta',-\alpha']$ on the real axis. Noting
$$K = B_1(t)\, \delta_0 - H(x,t), \tag{67}$$
theorem 4 is only saying that

$$\frac{1}{R_t(\xi) + 4\pi^2\xi^2 a_0(t)} = \phi_t(4\pi^2\xi^2) \tag{68}$$

if we note

$$k_t(z) = \frac{1}{\phi_t(z)} - a_0(t)z \tag{69}$$

we will have $R(\xi) = k_t(4\pi^2\xi^2)$ From 65 one has

$$k_t(z) = B_1 + \frac{B_1^2 - B_2}{a_0 z} + o(1/z) \tag{70}$$

But from ⑥⑥ one sees that

$$\begin{cases} \mathrm{Im}k_t(z) > 0 \quad \text{if} \quad \mathrm{Im}z > 0 & (71) \\[2mm] \text{Poles of } k_t \text{ are on the segment } [-\beta',-\alpha'] \text{ on the real axis so} \end{cases}$$

by a classical representation theorem there exists a positive measure $d\mu_t$ supported by the interval $[-\beta',-\alpha']$ such that

$$k_t(z) = B_1(t) - \int \frac{d\mu_t(\lambda)}{z - \lambda} \tag{72}$$

Then by inverse Fourier transform one finds

$$H(x,t) = \int \frac{1}{2\sqrt{|\lambda|}} \, e^{-\sqrt{|\lambda|}\,|x|} \, d\mu_t(\lambda) \tag{73}$$

So $H > 0$ (it must be so, as equation ⑥② must satisfy maximum principle); all
the different moments of $d\mu_t$ can be computed, theoretically, from ϕ , i.e.
from $a_0, B_1, \ldots B_m, \ldots$. Notice that $R(\xi) > 0$ and that its minimum value
corresponds to $z = 0$ and is $\dfrac{1}{\int \frac{d\nu t}{\lambda_2}} = \dfrac{1}{\text{weak lim} \frac{1}{b\epsilon(t)}}$.
Notice also that the nonlocal term comes from an exponentially decaying function, but of a very special type: the elementary solution of $-\dfrac{d^2}{dx^2} + c$.

   The analysis relies mainly on the fact that some function satisfies an
Herglotz type hypothesis ⑥⑥ , reminiscent of similar properties discussed by D.
Bergman in his talk.

Bibliography

The missing details for III are contained in

[1] F. Murat-L. Tartar: Calcul des variations et homogénéisation, Collection de
la Direction des Etudes et Recherches d'Electricité de France 57, Eyrolles,
Paris 1985, which is contained in the lecture notes of a summer school on
homogenization which, with the lectures of D. Bergman, J.L. Lions, G.
Papanicolaou and E. Sanchez Palencia, gives a good overview of the field.

The missing details for II are contained in

[2] L. Tartar: Estimations fines de coefficients homogénéisés, to appear in
Research Notes in Mathematics Pitman 1985 (Colloque De Giorgi, P. Kree ed.)

I referred in the text to many speakers at this workshop and thus to their written

contributions to this volume that will certainly contain more complete

bibliographical references.

VARIATIONAL ESTIMATES FOR THE OVERALL RESPONSE OF AN INHOMOGENEOUS NONLINEAR DIELECTRIC

J.R. Willis

School of Mathematics
Bath University
Bath   BA2 7AY
England

Abstract

For any problem that can be formulated as a "minimum energy" principle, a procedure is given for generating sets of upper and lower bounds for the energy. It makes use of "comparison bodies" whose energy functions may be easier to handle than those in the given problem. No structure for the energy functions is assumed in the formal development but useful results are most likely to follow when they are convex. When applied to linear field equations, the procedure yields the Hashin-Shtrikman variational principle, and so can be regarded as its generalization to nonlinear problems.

The procedure is applied explictly to a boundary value problem for an inhomogeneous, nonlinear dielectric. Then, a slight extension which describes randomly inhomogeneous media is applied, to develop bounds for the overall energy of a nonlinear composite, which reduce to the Hashin-Shtrikman bounds in the linear limit. Sample results are shown for a simple two-phase composite.

1. A Simple Observation

The objective of this work is to bound the total energy stored in an inhomogeneous body (such as a composite), when it is subjected to specified boundary conditions. For linear constitutive behaviour, use of a uniform "comparison" body and associated "polarizations" leads to a variational principle due to Hashin and Shtrikman (1962a, b, c; 1963) which has proved to be particularly effective for the construction of bounds. A review by Willis (1983) outlines some recent developments, and concludes by indicating a possible generalization that encompasses nonlinear behavior. This generalization is developed further here, and then applied in detail to the problem of bounding the energy in an inhomogeneous, nonlinear dielectric.

The simplicity of the basic idea is best exposed by first considering a general problem and specializing afterwards to electrostatics. Thus, for the moment, we consider the problem of minimizing a real-valued function $F(u)$, which is defined over some vector space B, when u is restricted to lie in a set $K \subset B$. Of course, for a minimizing u to exist, further structure needs to be specified. The simple argument to follow does not depend on it, however, so the problem will be relaxed to that of bounding the quantity $\inf_{u \in K} F(u)$.

Let $\overline{F}_0$ and $\underline{F}_0$ be defined over B and define

$$\overline{V}(v) = \sup_{u \in B} \{-\langle v,u\rangle + F(u) - \overline{F}_0(u)\} \, , \tag{1.1}$$

$$\underline{V}(v) = \inf_{u \in B} \{-\langle v,u\rangle + F(u) - \underline{F}_0(u)\} \, , \tag{1.2}$$

for $v \in B^*$, the vector space dual to B. (The possibilities that $\overline{V}(v) = \infty$ and $\underline{V}(v) = -\infty$ are not excluded, though then the observation to follow would be trivial.) It follows directly from (1.1) and (1.2) that

$$\langle v,u\rangle + \underline{F}_0(u) + \underline{V}(v) \leqslant F(u) \leqslant \langle v,u\rangle + \overline{F}_0(u) + \overline{V}(v) \tag{1.3}$$

for any $u \in B, v \in B^*$, and hence that

$$\inf_{u \in K} \{\langle v,u\rangle + \underline{F}_0(u) + \underline{V}(v)\} \leqslant \inf_{u \in K} F(u)$$

$$\leqslant \inf_{u \in K} \{\langle v,u\rangle + \overline{F}_0(u) + \overline{V}(v)\} \tag{1.4}$$

for any $v \in B^*$.

The remainder of this work develops implications of the inequalities (1.4). They are most useful, of course, when $\underline{F}_0, \overline{F}_0$ are such that the bounds are finite and equality can be attained for suitable choices of v. They provide, in fact, a concise and rather general statement of the variational principle of Hashin and Shtrikman (1962a, b, c; 1963): it will be seen below that the functions $\underline{F}_0$, $\overline{F}_0$ correspond to comparison media and that v corresponds to a polarization.

2. Electrostatics

The electric field E and potential ϕ are related so that

$$E = -\nabla \phi \tag{2.1}$$

and, in a dielectric, the electric displacement D is given as

$$D = \varepsilon E. \tag{2.2}$$

The properties of the dielectric are reflected in the second-order tensor ε, which is isotropic if the dielectric has isotropic structure. If the dielectric is nonlinear, then ε depends upon the value of the electric field $E(x)$, as well as on position x. The electrostatic problem that will be considered is defined by the equilibrium equation

$$\text{div } D = 0, \ x \in \Omega , \tag{2.3}$$

where D is related to E through (2.2) and hence to ϕ by (2.1). The body occupies the domain Ω and the potential ϕ is specified on the boundary, so that

$$\phi = \phi_0 , \ x \in \partial\Omega . \tag{2.4}$$

It will be assumed that an energy function $W(E,x)$ exists, so that (2.2) is equivalent to

$$D = \frac{\partial W}{\partial E} . \tag{2.5}$$

In the above, suffixes have been suppressed: in components, (2.5) would be written $D_i = \partial W / \partial E_i$, with corresponding expressions for the other equations.

By implication, $W(E,x)$ is (at least) once differentiable with respect to E and the problem defined by (2.1) to (2.4) has at least a weak solution (which is unique), if, in addition, $W(E,x)$ is strictly convex and coercive in the variable E. The space B can be taken as the set of vector fields whose components are square-integrable over Ω, and the subset K is the closed, convex set of elements of B which, in addition, have the representation (2.1) with ϕ satisfying the boundary condition (2.4). The weak solution is then the set of fields $D(x)$, $E(x)$, $\phi(x)$ which satisfy (2.1), (2.2) and (2.4) and for which $E(x)$ minimizes the energy

$$\int_\Omega W(E,x)dx.$$

The relevant theorem is well-known and is proved, for example, by Ekeland and Temam (1976).

Now choose a "comparison" body, with energy function $W_0(E)$; this function could in principle vary with x but this is not likely to be the case in practice. If D and E are related by (2.5), then also

$$D = \frac{\partial W_0}{\partial E} + P \quad , \tag{2.6}$$

if the "polarization" P satisfies

$$P = \frac{\partial W}{\partial E} - \frac{\partial W_0}{\partial E} \quad . \tag{2.7}$$

Equation (2.7) defines the classical electric polarization, if W_0 is chosen to characterize the properties of a vacuum; it is desirable here, however, to retain the option to define W_0 in any convenient way. Equation (2.3) now implies

$$div\left(\frac{\partial W_0}{\partial E}\right) + divP = 0, \tag{2.8}$$

so that the field E is generated in the comparison body by an appropriate distribution of charges. The field P, like D, belongs to the dual space B^*, which may be identified with B.

Now let $\overline{W}_0$, $\underline{W}_0$ be strictly convex and such that $\overline{W}_0 - W$ and $W - \underline{W}_0$ are strictly convex in E, for each x. Although weaker conditions would suffice, it is also natural to assume that $\overline{W}_0$, $\underline{W}_0$ are at least once differentiable. Now, in analogy with (1.1), (1.2), define

$$\overline{U}(P,x) = \inf_E \{P \cdot E - W(E,x) + \overline{W}_0(E)\}, \tag{2.9}$$

$$\underline{U}(P,x) = \sup_E \{P \cdot E - W(E,x) + \underline{W}_0(E)\}, \tag{2.10}$$

the infimum and supremum being taken pointwise, over $E \in R^3$. The reasoning that gave (1.4) now yields

$$\min_{E \in K} \int_\Omega \{P \cdot E + \underline{W}_0(E) - \underline{U}(P,x)\}\, dx$$

$$\leq \min_{E \in K} \int_\Omega W(E,x)\, dx$$

$$\leq \min_{E \in K} \int_\Omega \{P \cdot E + \overline{W}_0(E) - \overline{U}(P,x)\}\, dx \tag{2.11}$$

The correspondences between (2.11) and (1.4) are that

$$u \to E, \quad v \to P,$$

$$F(u) \to \int_\Omega W(E,x)\, dx,$$

$$V(v) \to -\int_\Omega U(P,x)\, dx$$

and so on (with bars being placed as appropriate). The sign of $U(P,x)$ is chosen to conform with notation already used by Willis (1983).

The minima that appear in the bounds are obtained by solving equations of the form of (2.8). The lower bound is maximized by choosing P so that

$$E = \frac{\partial \underline{U}}{\partial P}. \tag{2.12}$$

The differentiability of W and $\underline{W}_0$ ensures that $\underline{U}$ and $W - \underline{W}_0$ are Legendre duals, so that (2.12) implies (2.7) and the actual solution is realized; similar remarks apply to the upper bound.

In the particular case of a linear comparison material, so that

$$W_0(E) = \tfrac{1}{2}E\varepsilon_0E, \tag{2.13}$$

with ε_0 a constant tensor, (2.8) takes the form

$$\mathrm{div}(\varepsilon_0 E) + \mathrm{div}P = 0, \ x \in \Omega \tag{2.14}$$

and has solution

$$E = E_0 - \Gamma P, \tag{2.15}$$

where E_0 is the solution of (2.14) (with (2.1) and (2.4)) in the case $P = 0$, and Γ is a linear operator, closely related to the Green's function for (2.14); see, for example, Willis (1977). The bounds in (2.11) can then be put in the form

$$\int_\Omega \{\tfrac{1}{2}E_0\varepsilon_0E_0 + P.E_0 - \tfrac{1}{2}P\Gamma P - U(P,x)\} \ dx, \tag{2.16}$$

upon use of the easily-derived properties

$$\Gamma\varepsilon_0\Gamma = \Gamma \ , \ \int_\Omega E_0\varepsilon_0\Gamma P \ dx = 0. \tag{2.17}$$

Since the bounds can be attained by suitable choices of P, the original problem is equivalently characterized by the stationary principle

$$\delta \int_\Omega \{P.E_0 - \tfrac{1}{2}P\Gamma P - U(P,x)\} \ dx = 0. \tag{2.18}$$

In the case of elasticity, this stationary principle was derived directly by Willis (1983), with U defined as the Legendre dual of $(W - W_0)$; the demonstration that it yields bounds has not been given previously. The classical Hashin-Shtrikman principle is obtained by specializing to linear material behaviour, so that $W(E)$ takes the form

$$W(E,x) = \tfrac{1}{2} E \varepsilon E, \tag{2.19}$$

with the tensor ε depending on x only. In this case, either of (2.9) or (2.10) yield

$$U(P,x) = \tfrac{1}{2}P(\varepsilon - \varepsilon_0)^{-1}P. \tag{2.20}$$

Then, the expression (2.16) yields explicit upper and lower bounds, in line with (2.11), when ε_0 is chosen as $\overline{\varepsilon}_0$ or $\underline{\varepsilon}_0$ to give $\overline{U}$ or $\underline{U}$.

The approach developed in Section 1 can also be applied, taking the complementary energy principle as starting point. This leads to bounds on the energy which are similar in general form to (2.11) but employ different variables. These bounds, and their relationship with (2.11), will be discussed elsewhere.

3. Application to a Composite

Consider now a medium which is composed of n distinct constituents, or phases, with energy functions $W_1(E)$, $W_2(E)$, ... $W_n(E)$ say, so that

$$W(E,x) = \sum_{r=1}^{n} W_r(E)f_r(x), \tag{3.1}$$

where $f_r(x)$ takes the value 1 if x lies in material of rth type and zero otherwise. For convenience of description, such a medium is called a composite. Randomly inhomogeneous media will be considered, which means that the functions f_r will be taken as random fields. Correspondingly, bounds will be sought for the ensemble mean of the energy, rather than the energy in any particular realization of the composite. The inequalities (2.11) survive ensemble averaging; hence, bounds will be sought by specializing the inequalities

$$\left< \min_{E \in K} \int_{\Omega} \{P \cdot E + \underline{W}_0(E) - \underline{U}(P,x)\}dx \right>$$

$$\leq \left< \min_{E \in K} \int_{\Omega} W(E,x)dx \right> \leq \left< \min_{E \in K} \int_{\Omega} \{P \cdot E + \overline{W}_0(E) - \overline{U}(P,x)\} dx \right> \tag{3.2}$$

to particular comparison media defined by $\underline{W}_0$, $\overline{W}_0$, and to particular configuration-dependent polarizations $P(x)$. Corresponding to the form (3.1), we have

$$U(P,x) = \sum_{r=1}^{n} U_r(P)f_r(x), \tag{3.3}$$

where $U_r(P)$ is the Legendre dual of $W_r - W_0$ (with bars placed as appropriate). The simplest non-trivial polarization $P(x)$ has a similar form:

$$P(x) = \sum_{r=1}^{n} P_r(x)f_r(x), \tag{3.4}$$

in which the functions $P_r(x)$ are deterministic. The best bounds that can be obtained from the class of polarizations (3.4) are then found by choosing the functions $P_r(x)$ optimally.

Having chosen $P(x)$, it is necessary to solve a boundary value problem of the form (2.8). This is likely to be tractable only if the comparison medium is linear, so that W_0 is a quadratic function. It is, of course not guaranteed that quadratic functions $\overline{W}_0, \underline{W}_0$ can be chosen so that $\overline{W}_0 - W$, $W - \underline{W}_0$ are strictly convex. Our approach, however, will be to proceed with a quadratic W_0 and consider afterwards the conditions under which upper and/or lower bounds can be obtained in this way. When W_0 is quadratic, E has the representation (2.15), which may be given in the slightly more explicit form

$$E(x) = E_0(x) - \int_\Omega dx'\ \Gamma(x,x')P(x')\ , \tag{3.5}$$

showing Γ as an integral operator (in the sense of generalized functions) with kernel $\Gamma(x,x')$. When $P(x)$ is given by (3.4), the bounds in (3.2) take the form

$$< \int_\Omega dx\ \{\tfrac{1}{2} E_0 \varepsilon_0 E_0 + \sum_r P_r \cdot E_0 f_r(x) - \tfrac{1}{2} \sum_r \sum_s P_r f_r(x) \int_\Omega dx' \Gamma(x,x') f_s(x') P_s$$

$$- \sum_r U_r(P_r)f_r(x)\}> , \tag{3.6}$$

in line with the simplification (2.16). With the definitions

$$p_r(x) = \langle f_r(x) \rangle \quad , \quad p_{rs}(x,x') = \langle f_r(x) f_s(x') \rangle \quad , \tag{3.7}$$

this becomes

$$\int_\Omega dx \{ \tfrac{1}{2} E_0 \epsilon_0 E_0 + \sum_r P_r \cdot E_0 P_r (x) - \tfrac{1}{2} \sum_r \sum_s P_r \int_\Omega dx' \Gamma(x,x') p_{rs}(x,x') P_s$$

$$- \sum_r U_r(P_r) p_r(x) \} \tag{3.8}$$

and (3.8) is optimized (that is, maximized for the lower bound or minimized for the upper bound) when the functions $P_r(x)$ satisfy

$$\frac{\partial U_r}{\partial P_r} P_r + \sum_s \int_\Omega dx' \ \Gamma(x,x') p_{rs}(x,x') P_s(x') = E_0 P_r. \tag{3.9}$$

It is convenient to ensemble average (3.5), so that

$$\langle E \rangle(x) = E_0(x) - \sum_s \int_\Omega dx' \Gamma(x,x') p_s(x') \ P_s(x') \quad , \tag{3.10}$$

and to combine (3.9), (3.10) to give

$$\frac{\partial U_r}{\partial P_r} P_r + \sum_s \int_\Omega dx' \ \Gamma(x,x')(p_{rs}(x,x') - p_r(x) p_s(x')) P_s(x') = \langle E \rangle p_r \ . \tag{3.11}$$

When the functions P_r satisfy (3.10), (3.11), the bounds (3.2) can be expressed in terms of $\langle E \rangle$ rather than E_0 (although these are related by (3.10)). Elementary algebra gives

$$\langle P \cdot E + W_0(E) - U(P) \rangle$$

$$= \tfrac{1}{2} \langle E \rangle \ \epsilon_0 \ \langle E \rangle + \tfrac{1}{2} \sum_r p_r \{ P_r \cdot \langle E \rangle + (P_r \cdot \frac{\partial U_r}{\partial P_r} - 2U_r) \} \ . \tag{3.12}$$

Suppose now that the composite has no long-range order, so that $p_{rs}(x,x') - p_r(x) p_s(x')$ approaches zero when $|x - x'|$ is large: expressed loosely, this means that $p_{rs} - p_r p_s$ is negligible when $|x-x'| > \ell$ where ℓ is a "correlation length", which will be assumed small compared with a typical macroscopic dimension, L, of Ω. If boundary conditions are applied for which E_0 varies slowly relative to the microscale ℓ, it is credible that P_r likewise varies slowly, so that equations (3.11) reduce, in the limit $\ell/L \to 0$, to the algebraic set

$$\frac{\partial U_r}{\partial P_r} + \sum_s A_{rs} \, P_s(x) = \,<E>(x) \,, \tag{3.13}$$

where

$$P_r(x) A_{rs}(x) \;=\; \int_\Omega dx' \Gamma(x,x')(P_{rs}(x,x') - P_r(x)P_s(x')). \tag{3.14}$$

Then, $P_r(x)$ is expressible in terms of $<E>(x)$ and the bounds (3.2) take the form

$$\int_\Omega \widetilde{W}(<E>(x)) \, dx,$$

where

$$\widetilde{W}(<E>) = \tfrac{1}{2} <E> \, \varepsilon_0 \, <E>$$
$$+ \frac{1}{2} \sum_r \, P_r \{ P_r \cdot <E> + (P_r \cdot \frac{\partial U_r}{\partial P_r} - 2U_r) \} \,. \tag{3.15}$$

In the particular case of a linear composite (so that W_r and U_r are quadratic functions), the bounds outlined above are precisely the Hashin-Shtrikman bounds, as generalized by Willis (1977). They reduce to the classical Hashin-Shtrikman bounds when the functions $P_{rs}(x,x')$ are translation-invariant and isotropic, and so depend only upon $|x - x'|$. More generally, $\widetilde{W}(<E>)$ yields estimates for the energy in the composite, which in many cases will be bounds, and which utilize statistical information on the composite through the one- and two-point probabilities P_r, P_{rs} .

4. A Particular Isotropic Composite

Consider a composite dielectric whose rth phase has constitutive relation

$$D = (\epsilon_r + \gamma_r |E|^2)E, \qquad (4.1)$$

where ϵ_r , γ_r are scalars, so that

$$W_r(E) = \tfrac{1}{2} \epsilon_r |E|^2 + \tfrac{1}{4} \gamma_r |E|^4. \qquad (4.2)$$

The composite will be assumed to be statistically uniform, so that p_r is independent of x and equal to the volume concentration of the rth phase, and isotro-
pic (at least for two-point statistics) so that p_{rs} depends upon $|x-x'|$ only.
A particular case of this type of composite was considered by Miksis (1983); he
applied a "self-consistent" procedure to a composite consisting of a linear matrix
containing an isotropic distribution of spherical inclusions, whose behaviour was
described by (4.1).

For this example, the comparison medium is taken to be isotropic, so that

$$W_0(E) = \tfrac{1}{2} \epsilon_0 |E|^2 .$$

Correspondingly,

$$U_r(P_r) = P_r \cdot E_r - \tfrac{1}{2}(\epsilon_r - \epsilon_0)|E_r|^2 - \tfrac{1}{4} \gamma_r |E_r|^4 , \qquad (4.4)$$

where E_r is related to P_r so that

$$P_r = (\epsilon_r - \epsilon_0 + \gamma_r |E_r|^2)E_r . \qquad (4.5)$$

An equivalent statement is that

$$E_r = \frac{\partial U_r}{\partial P_r} ; \qquad (4.6)$$

this is useful for substitution into (3.11). The solution of (4.5) for E_r in
terms of P_r , can be given in the form

$$E_r = \frac{P_r}{(\varepsilon_r - \varepsilon_0)} \; G\left(\frac{\gamma_r^{1/2}|P_r|}{(\varepsilon_r - \varepsilon_0)^{3/2}}\right) , \qquad (4.7)$$

where the function G is defined so that $y = x\,G(x)$ is the (unique) real solution of

$$y^3 + y = x ; \qquad (4.8)$$

equivalently, G satisfies

$$(1 + x^2 G^2)G = 1. \qquad (4.9)$$

Equations (3.11) become explicit once the constants A_{rs} are determined. These have already been found in the context of linear problems: it follows from work of Willis (1977), for example, that

$$A_{rs} = \frac{1}{3\varepsilon_0} \; (P_r \, \delta_{rs} - P_r P_s) \quad \text{(no sum on r).} \qquad (4.10)$$

Equations (3.11) therefore become

$$E_r + \frac{1}{3\varepsilon_0} \; (P_r - \sum_s P_s P_s) = \; <E> , \qquad (4.11)$$

where E_r, P_r are related by (4.5), (4.6) or (4.7). Then $\widetilde{W}(<E>)$ can be expressed in the form

$$\widetilde{W}(<E>) = \tfrac{1}{2} \varepsilon_0 \, |<E>|^2$$

$$+ \frac{1}{2} \sum_r P_r \; \{[(\varepsilon_r - \varepsilon_0) + \gamma_r |E_r|^2] E_r \cdot <E> - \tfrac{1}{2} \gamma_r |E_r|^4\} . \qquad (4.12)$$

This completes the formal prescription for generating the estimate $\widetilde{W}(<E>)$. It is still necessary to solve the system (4.11), with P_r given in terms of E_r by (4.5). This problem is taken up in one particular case in the section that follows.

An Isotropic Two-Phase Composite

For the sake of simple illustration, consider a composite consisting of just two phases. The first, for which $r=1$, has constitutive relation (4.1) and energy function (4.2), while the second is linear, with dielectric constant ε_2. The dielectric constant ε_0 of the comparison body is taken equal to ε_2. Although $W_2 - W_0$ is then not strictly convex (being precisely zero), a lower bound for the energy in the composite is still obtained, in the case $\varepsilon_2 < \varepsilon_1$, by taking $P_2 = 0$. The opposite case, $\varepsilon_2 > \varepsilon_1$, would generate an upper bound for the energy, at least when $\langle E \rangle$ is sufficiently small, but this will not be examined here.

The equations that define E_1, E_2 are now

$$\{1 + \frac{p_2}{3\varepsilon_2} \ (\varepsilon_1 - \varepsilon_2 + \gamma_1 |E_1|^2)\} \ E_1 = \langle E \rangle, \tag{4.13}$$

$$p_1 \ E_1 + p_2 \ E_2 = \langle E \rangle , \tag{4.14}$$

with $p_1 + p_2 = 1$. The solution of (4.13) is

$$E_1 = \frac{3\varepsilon_2 \langle E \rangle}{3\varepsilon_2 + p_2(\varepsilon_1 - \varepsilon_2)} \ G \ \{ \frac{(p_2\gamma_1)^{1/2} 3\varepsilon_2 |\langle E \rangle|}{[3\varepsilon_2 + p_2(\varepsilon_1 - \varepsilon_2)]^{3/2}} \} . \tag{4.15}$$

Elementary manipulation then gives

$$\tilde{W}(\langle E \rangle) = \frac{1}{2} \ [\ \varepsilon_2 + \frac{p_1 \ (\varepsilon_1 - \varepsilon_2)(3\varepsilon_2)}{3\varepsilon_2 + p_2(\varepsilon_1 - \varepsilon_2)}$$

$$+ \frac{p_1(3\varepsilon_2)^2 (1-G)(1-G/2)}{p_2[3\varepsilon_2 + p_2(\varepsilon_1 - \varepsilon_2)]} \ ]|\langle E \rangle|^2 , \tag{4.16}$$

where G has the argument shown in (4.15).

Several limiting cases are of interest, at least as a check on (4.16). First, if $|\langle E \rangle|$ is small, $G \sim 1$ and (4.16) reduces to the corresponding linear Hashin-Shtrikman bound. Next, if $|\langle E \rangle|$ is large, $G \rightarrow 0$ and (4.16) gives

$$\widetilde{W}(<E>) \sim \frac{1}{2} \left[\varepsilon_2 + \frac{p_1}{p_2} (3\varepsilon_2) \right] |<E>|^2 \tag{4.17}$$

which is independent of ε_1. The same expression follows from the linear Hashin-Shtrikman bound in the limit $\varepsilon_1 \to \infty$. This is not surprising since the "dielectric constant" of phase 1 tends to infinity as $|<E>|$ tends to infinity. Finally, when p_1 is small, (4.16) gives, to first order in p_1,

$$\widetilde{W}(<E>) \sim \frac{1}{2} \left[\varepsilon_2 + \frac{p_1(\varepsilon_1 - \varepsilon_2)(3\varepsilon_2)}{\varepsilon_1 + 2\varepsilon_2} + \frac{p_1(3\varepsilon_2)^2}{\varepsilon_1 + 2\varepsilon_1} (1 - G)(1 - G/2) \right] |<E>|^2 .$$

$$\tag{4.18}$$

Details are not given, but this result can be obtained independently by solving the problem of a single nonlinear spherical inclusion in a linear matrix. The estimate (4.16) is therefore exact, to first order in p_1, for a matrix containing a dilute suspension of nonlinear spheres. In this case, the direct solution also gives the relation

$$<D> = \widetilde{\varepsilon}<E>, \tag{4.19}$$

where

$$\widetilde{\varepsilon} = \varepsilon_2 + \frac{p_1(\varepsilon_1 - \varepsilon_2)(3\varepsilon_2)}{\varepsilon_1 + 2\varepsilon_2} + \frac{p_1(3\varepsilon_2)^2}{\varepsilon_1 + 2\varepsilon_2}(1 - G) . \tag{4.20}$$

It can be checked that (4.18) and (4.20) are consistent with the overall relation

$$<D> = \partial\widetilde{W}/\partial<E>, \tag{4.21}$$

by noting that

$$\int (1 - G) \, x dx = \frac{x^2}{2} (1-G)(1 - G/2). \tag{4.22}$$

This follows by writing $y = x\, G(x)$, integrating by parts and using the relation (4.8).

More generally, this observation can be exploited to yield

$$\frac{\partial \tilde{W}}{\partial \langle E \rangle} = \tilde{\varepsilon} \langle E \rangle, \tag{4.23}$$

where

$$\tilde{\varepsilon} = \varepsilon_2 + \frac{p_1(\varepsilon_1 - \varepsilon_2)(3\varepsilon_2)}{3\varepsilon_2 + p_2(\varepsilon_1 - \varepsilon_2)} + \frac{p_1(3\varepsilon_2)^2(1-G)}{p_2[3\varepsilon_2 + p_2(\varepsilon_1 - \varepsilon_2)]} \quad . \tag{4.24}$$

It should be noted, however that, while $\tilde{W}(\langle E \rangle)$ bounds the energy from below when $\varepsilon_2 < \varepsilon_1$, $\tilde{\varepsilon}$ is just an estimate for the overall "nonlinear dielectric constant" which is guaranteed to be a lower bound in the linear limit $\langle E \rangle \to 0$.

Some sample results are displayed in the figures. Figure 1 shows plots of $W/(\frac{1}{2} \varepsilon_2 |\langle E \rangle|^2)$ and $\tilde{\varepsilon}/\varepsilon_2$ against $(\gamma_1/\varepsilon_2)^{1/2}|\langle E \rangle|$, for the case $\varepsilon_1 = 2\varepsilon_2$, at the two concentrations $p_1 = 0.3, 0.7$. Figure 2 shows similar plots, except that $\varepsilon_1 = \varepsilon_2$, so that the nonlinear term is solely responsible for the inhomogeneity. Miksis (1983) studied this latter example (in the case of nonlinear spheres distributed in a linear matrix) by a "self-consistent" procedure which made no explicit allowance for pairwise statistics. The results that are shown are unlikely to have great practical importance; their main interest is that they provide a certain amount of "honest" information about the overall energy function, even though its functional form is unknown. A more complete study will be presented elsewhere.

References

Ekeland, I. and Temam, R., "Convex Analysis and Variational Problems". North Holland, Amsterdam (1976).

Hashin, Z. and Shtrikman, S., "On some variational principles in anisotropic and nonhomogeneous elasticity", J. Mech. Phys. Solids 10, 335-342 (1962a).

Hashin, Z. and Shtrikman, S., "A variational approach to the theory of the elastic behaviour of polycrystals", J. Mech. Phys. Solids 10, 343-350 (1962b).

Hashin, Z. and Shtrikman, S., "A variational approach to the theory of the magnetic permeability of multiphase materials", J. Appl. Phys. 33, 3125-3131 (1962c).

Hashin, Z. and Shtrikman, S., "A variational approach to the theory of the elastic behaviour of multiphase materials", J. Mech. Phys. Solids 11, 127-140 (1963).

Miksis, M.J., "Dielectric constant of a nonlinear composite material", SIAM J. Appl. Math. 43, 1140-1155 (1983).

Willis, J.R., "Bounds and self-consistent estimates for the overall properties of anisotropic composites", J. Mech. Phys. Solids 25, 185-202 (1977).

Willis, J.R., "The overall elastic response of composite materials", J. Appl. Mech. 50, 1202-1209 (1983).

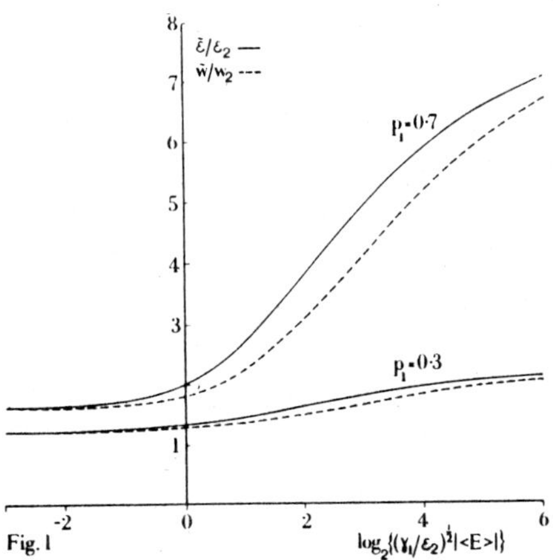

Fig 1. Plots of $\tilde{\varepsilon}/\varepsilon_2$ and $\tilde{W}/W_2$ (where $W_2 = \frac{1}{2}\varepsilon_2|<E>|^2$) against $(\gamma_1/\varepsilon_2)^{1/2}|<E>|$, with $\varepsilon_1 = 2\varepsilon_2$, for concentrations $p_1 = 0.3$ and $p_1 = 0.7$. The curves for $\tilde{W}$ provide lower bounds for the energy, but those for $\tilde{\varepsilon}$ may not be bounds for its derivative. If $\tilde{W}$ were quadratic, the plots for $\tilde{\varepsilon}/\varepsilon_2$ and $\tilde{W}/W_2$ would coincide. They tend to coincide for small and for large $|<E>|$.

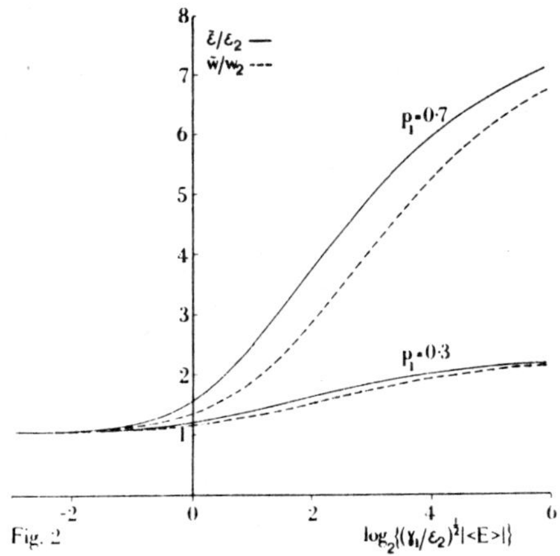

Fig 2. As Figure 1, except that $\varepsilon_1 = \varepsilon_2$ so that the term involving γ_1 is solely responsible for the inhomogeneity.

Table of Contents from Other Volumes
from the Program in
Continuum Physics and Partial Differential Equations

Theory and applications of liquid crystals

January 21 – January 25, 1985

J. L. Ericksen
D. Kinderlehrer
Conference Committee

Tentative contributors: Berry, G., Brezis, H., Capriz, G., Choi, H. I.,
Cladis, P., Di Benedetto, E., Gulliver, R., Hardt. R. and Kinderlehrer, D.,
Leslie, F., Miranda, M., Ryskin, G., Sethna, J., and Spruck, J.

Amorphous polymers and non-newtonian fluids

March 4 – March 8, 1985

J. L. Ericksen
D. Kinderlehrer
M. Tirrell
S. Prager
Conference Committee

Tentative contributors: Bird, R., Caswell, B., Dafermos, C., Hrusa,
W. and Renardy, M., Joseph, D. D., Kearsley, E., Marcus, M. and Mizel, V.,
Nohel, J. and Renardy, M., Rabin, M., and Tirrell, M.

Oscillation theory, computation, and
methods of compensated compactness

April 1 – April 4, 1985

C. Dafermos
J. L. Ericksen
D. Kinderlehrer
M. Slemrod
Conference Committee

Chacon, T. and Pironneau, O. — Convection of microstructures by incompressible and slightly compressible flows

Colella, P. — Numerical calculation of fluid flows with strong shocks

DiPerna, R. — Oscillations in solutions to nonlinear differential equations

Forest, M.G. — Geometry and modulation theory for the periodic nonlinear Schroedinger equation

Harten, A. — On high-order accurate interpolation for non-oscillatory shock capturing schemes

Lax, P. — Dispersive difference equations

Majda, A. — Nonlinear geometric optics for hyperbolic systems of conservation laws

McLaughlin, D. — On the construction of a modulating multiphase wavetrain for a perturbed KdV equation

Nunziato, J., Gartling, D., and Kipp, M. — Evidence of nonuniqueness and oscillatory solutions in in computational fluid mechanics

Osher, S. — Very high order accurate TVD schemes

Rascle, M. — Convergence of approximate solutions to some systems of conservation laws: a conjecture on the product of the Riemann invariants

Oscillation theory, computation, and methods of compensated compactness

Schonbek, M. — Applications of the theory of compensated compactness

Serre, D. — A general study of the commutation relation given by L. Tartar

Slemrod, M. — Interrelationships among mechanics, numerical analysis, compensated compactness, and oscillation theory

Venakides, S. — The solution of completely integrable systems in the continuum limit of the spectral data

Warming, R. — Stability of finite-difference approximations for hyperbolic initial boundary value problems

Yee, H. — Construction of a class of symmetric TVD schemes

(tentative contents)

Metastability and incompletely posed problems

May 6 - May 10, 1985

S. Antman
J. L. Ericksen
D. Kinderlehrer
I. Müller
Conference Committee

Antman, S.	Dissipative mechanisms
Ball, J.	Does rank-one convexity imply quasiconvexity?
Brezis, H.	Metastable harmonic maps
Calderer, M.	Bifurcation of constrained problems in thermoelasticity
Chipot, M. and Luskin,M.	The compressible Reynolds' lubrication equation
Ericksen, J.	Twinning of crystals I
Evans, L. C.	Quasiconvexity and partial regularity in the calculus of variations
Goldenfeld, N.	Introduction to pattern selection in dendritic solidification
Gurtin, M.	Some results and conjectures in the gradient theory of phase transitions
James, R.	The stability and metastability of quartz
Kenig, C.	Continuation theorems for Schrodinger operators
Kinderlehrer, D.	Twinning of crystals II
Lions, J. L.	Asymptotic problems in distributed systems

Metastability and incompletely posed problems

Liu, T.P.	Stability of nonlinear waves
Mosco, U.	Variational stability and relaxed Dirichlet problems
Müller, I.	Simulation of pseudoelastic behaviour in a system of rubber balloons
Pitteri, M.	A contribution to the description of natural states for elastic crystalline solids
Rogers, R.	Nonlocal problems in electromagnetism
Salsa, S.	The Nash-Moser technique for an inverse problem in potential theory related to geodesy
Vazquez, J.	Hyperbolic aspects in the theory of the porous medium equation
Vergara-Caffarelli, G.	Green's formulas for linearized problems with live loads
Wright, T.	Some aspects of adiabatic shear bands

(tentative contents)

Dynamical problems in continuum physics

June 3 – June 7, 1985

J. Bona
C. Dafermos
J. L. Ericksen
D. Kinderlehrer
Conference Committee

Beals, M.   Presence and absence of weak singularities in nonlinear waves

Beatty, M.   Some dynamical problems in continuum physics

Beirao da Veiga, H.   Existence and asymptotic behavior for strong solutions of the Navier Stokes equations in the whole space

Bell, J.   A confluence of experiment and theory for waves of finite strain in the solid continuum

Bona, J.   Shallow water waves and sediment transport

Chen, P.   Classical piezoelectricity: is the theory complete?

Keller, J.   Acoustoelasticity

McCarthy, M.   One dimensional finite amplitude pulse propagation in electroelastic semiconductors

Müller, C.   Extended thermodynamics of ideal gases

Pego, R.   Phase transitions in one dimensional nonlinear viscoelasticity: admissibility and stability

Shatah, J.   Recent advances in nonlinear wave equations

Slemrod, M.   Dynamic phase transitions and compensated compactness

Spagnolo, S.   Some existence, uniqueness, and non-uniqueness results for weakly hyperbolic equations in Gevrey classes

Strauss, W.   On the dynamics of a collisionless plasma

(tentative contents)